W9-CIK-089

Environmental Quality
and Residuals Management

Environmental Quality and Residuals Management

Report of a Research Program on Economic, Technological, and Institutional Aspects

By ALLEN V. KNEESE and BLAIR T. BOWER

Published for Resources for the Future
By The Johns Hopkins University Press
Baltimore and London

Copyright © 1979 by Resources for the Future, Inc.
All rights reserved
Manufactured in the United States of America
Library of Congress Catalog Card Number 79-2181
ISBN 0-8018-2245-9
ISBN 0-8018-2286-6, paperback
Library of Congress Cataloging in Publication Data will be found
on the last printed page of this book.

RESOURCES FOR THE FUTURE, INC.
1755 Massachusetts Avenue, N.W., Washington, D.C. 20036

Board of Directors:
Gilbert F. White, *Chairman,* Irving Bluestone, Harrison Brown, Roberto de
O. Campos, Anne P. Carter, Emery N. Castle, William T. Coleman, Jr., F.
Kenneth Hare, Charles J. Hitch, Franklin A. Lindsay, Charles F. Luce, George
C. McGhee, Ian MacGregor, Vincent E. McKelvey, Laurence I. Moss, Frank
Pace, Jr., Stanley H. Ruttenberg, Lauren Soth, Janez Stanovnik, Russell E.
Train, M. Gordon Wolman.

Honorary Directors: Horace M. Albright, Erwin D. Canham, Edward J.
Cleary, Hugh L. Keenleyside, Edward S. Mason, William S. Paley, John W.
Vanderwilt

President: Emery N. Castle

Secretary-Treasurer: John E. Herbert

Resources for the Future is a nonprofit organization for research and education in the development, conservation, and use of natural resources and the improvement of the quality of the environment. It was established in 1952 with the cooperation of the Ford Foundation. Grants for research are accepted from government and private sources only if they meet the conditions of a policy established by the Board of Directors of Resources for the Future. The policy states that RFF shall be solely responsible for the conduct of the research and free to make the research results available to the public. Part of the work of Resources for the Future is carried out by its resident staff; part is supported by grants to universities and other nonprofit organizations. Unless otherwise stated, interpretations and conclusions in RFF publications are those of the authors; the organization takes responsibility for the selection of significant subjects for study, the competence of the researchers, and their freedom of inquiry.

This book is a product of the Quality of the Environment Division, which is directed by Walter O. Spofford, Jr. It was edited by Sally A. Skillings.

RFF editors: Joan R. Tron, Ruth B. Haas, Jo Hinkel, Sally A. Skillings

Contents

Text Tables

Appendix Table

Text Figures

Appendix Figures

Preface

This brief preface has three purposes. The first is to indicate the range of applicability of the methodologies and analytical approaches discussed in this book. The second is to emphasize that this report is the joint product of many researchers. The third is to acknowledge the debt owed by the principal authors to others who helped us.

Since the program of research on which this report is based was first undertaken, efforts to cope with residuals problems have increased in many, if not most, of the countries of the world. The point of departure for the economic aspects of our research is market theory. But the fundamental problems of managing common property resources and of developing and applying implementation incentives to induce use of those resources consistent with socially determined goals are common to all political-economic systems and to all levels of economic development. Thus, the analytical approaches and quantitative models discussed here are as applicable to nonmarket as to market economies. Our experiences over the last decade have served to make that fact quite apparent (as appeared in the RFF Research Paper R-11 of Daniel J. Basta, James L. Lounsbury, and Blair T. Bower, entitled *Analysis for Residuals–Environmental Quality Management,* 1978). However, this is not to say that modeling and analysis should be highly sophisticated everywhere. On the contrary, in many cases in any economy, simple models are sufficient to provide the requisite information for decisions.

With respect to the second purpose, it will be apparent throughout that this book is the joint product of a considerable number of individuals, heralded and unheralded. It would not have been possible without the various researchers whose work comprises its raw material. Some of them have also prepared material specifically for inclusion in it, as we note at appropriate points.

In addition, we have two other categories of debts to acknowledge. First, this report would not have come to fruition without the unflagging efforts of Vera Ullrich, who not only typed various versions of the manuscript, but managed to maintain three-way communication in spite of our various comings and goings. Pathana Thananart produced many of the original charts, graphs, and flow diagrams. Second, the report has been substantially improved by the helpful comments of reviewers, whose indulgence we asked where we did not accept their suggestions in toto. Clifford Russell, Ralph Luken, Nathaniel Wollman, and Charles Ehler made extensive comments on the entire manuscript. Walter Spofford, Edwin Haefele, Maynard Hufschmidt, and Henry Peskin commented on specific chapters.

Finally, we wish to acknowledge our debt to Sally Skillings who did an excellent job of editing a long and involved manuscript.

Washington, D.C.   Allen V. Kneese
February 1979   Blair T. Bower

PART ONE

Background

CHAPTER 1

The Residuals Problem

General Nature

Environmental pollution has been a problem for human beings from the time they began to live together in large numbers in cities.[1] But the problem has certainly changed over time and is becoming in many ways more complex. In developed countries, environmental conditions that directly and immediately affect the daily lives of the mass of citizens improved, if somewhat erratically, at least until the last quarter of the twentieth century. What, then, has happened to bring environmental, especially environmental pollution, problems into such prominence? In the developed nations, with which this book is primarily concerned, at least three important tendencies can be identified.

First, the growth in industrial production, energy conversion, and transport of goods and people is an important factor. These activities have reached levels at which the associated flows of materials and energy from concentrated states in nature to degraded and diluted states in the environment have begun to alter the physical, chemical, and biological quality of the atmosphere in local areas, regions, and, in a few cases, the entire

[1] Horror stories about environmental conditions in medieval and even relatively modern times can easily be collected. The situation in London is especially well documented. In the fourteenth century, butchers had been assigned a spot at Seacoal Lane near Fleet prison. A royal document about this reads, "By the killing of great beasts, from whose putrid blood running down the streets and the bowels cast into the Thames, the air in the city is very much corrupted and infected, whence abominable and most filthy stinks proceed, sickness and many other evils have happened to such as have abode in the said city, or have resorted to it." Quoted in B. Lambert, *History and Survey of London,* vol. 1 (London, 1806) p. 241. Five centuries later, Charles Dickens is reported to have been moved to say of London, "He knew of many places in it unsurpassed in the accumulated horrors of their long neglect by the dirtiest old spots in the dirtiest old towns, under the worst old government of Europe." This statement is from *The Public Health as a Public Question: First Report of the Metropolitan Sanitary Association,* Address of Charles Dickens, Esq., London, 1850.

3

globe. Furthermore, it is now possible to detect even small changes in the quality characteristics of these large natural systems.

Second, a great variety of exotic materials is being inserted into the environment. Modern chemistry and physics have subjected the world's ecological systems to strange substances to which they cannot adapt, at least not quickly, or for which adaptation is highly specific among species and therefore disruptive.

Third, ordinary people in developed countries have come to expect standards of cleanliness, safety, and healthfulness in their surroundings that only the upper classes enjoyed in earlier times. Clearly, growing demand for higher quality environment is as much a factor in recent concern as actual deterioration in environmental quality. Indeed, in some cases concern has risen at the same time that environmental conditions have actually improved.

These are very large and complex matters. But some fairly simple concepts, derived from general economic theory of resources allocation, provide considerable understanding of the root causes of the present residuals problem in economies like that of the United States. To describe these causes in a phrase, they result from "failures in our system of economic incentives." Before elaborating on this simple statement, it is necessary to say a little about how economic theory visualizes the functioning of the economy.[2]

The market system is on the one hand an actual functioning economic system and on the other an intellectual idealization or a "model." This abstract model has been developed and analyzed over the decades by economists who were curious about how an apparently so uncoordinated set of activities seemed to achieve a rather impressive degree of economic order. They also wondered whether this order had desirable properties or was just orderly. The market model was shown to have desirable properties if a certain value premise is accepted and if the economy has certain structural characteristics.

The value premise is that the personal wants of the individuals in the society should guide the use of resources in production, distribution, and exchange, and that those personal wants can most efficiently be met

---

[2] It is also relevant to point out that environmental pollution problems in nonmarket economies are similarly a result of inadequate incentives—political, social, and economic. Many of the conclusions presented in this book concerning resource allocation and its relationship to residuals problems are equally relevant to nonmarket economies.

through the seeking of maximum profits by all producers. The necessary structural assumptions are that: (1) all markets are competitive—by this is meant that no particular firm or individual can affect any market price significantly by decreasing or increasing the supply of goods and services offered; (2) all participants in the market are fully informed as to the quantitative and qualitative characteristics of goods and services and the terms of exchange; (3) the distribution of income and wealth in the society is governed by rules that the society has collectively determined to be ethically acceptable, and, usually implicit; and (4) all assets can be individually owned and managed without violating the competitive assumption.

Also implicit in this model of the market system is the assumption that where choices must be made outside the market, there is a political mechanism through which collective choices are made effectively and efficiently. If all these conditions hold, it can be concluded that the best social solution to the resources use problem—one on which everyone could in principle agree—is to limit the role of government to deciding questions of income distribution, providing rules of property and exchange, and enforcing competition, and letting the market take care of the rest.

The connection between this normative intellectual idealization and the actual working economy and political structure has always been tenuous at best. But the model has served as a criterion against which the market economy can be judged as a social device for the pursuit of economic welfare. Thus it has provided strong intellectual support for public measures to sustain competition and improve the honesty and equity of the economic system. Throughout this book, although often only obliquely, this model is used as a reference point. The ways in which the real economy departs from this idealization are often a useful starting point for the consideration of public policy issues.[3]

Before considering one of these departures of special pertinence to residuals management questions, it is useful to remind ourselves of a simple principle of physics. When materials—minerals, fuels, gases, and organic materials—are extracted and harvested from nature and used by producers and so-called consumers, their mass is not altered in these

[3] No attempt is made here to catalogue and discuss all departures, that is, monopolistic and oligopolistic situations, lack of knowledge, and so forth. This has been ably done elsewhere. Only those departures of direct relevance to residuals–environmental quality problems are discussed herein.

processes, except in trivial amounts. Materials and energy residuals are generated in production and consumption activities, and the mass of the former must be about equal to that initially extracted from nature.[4] Accordingly, it is basically deceptive to speak of the consumption of goods.[5]

Material objects yield services that are used for greater or lesser lengths of time. But when they no longer yield services, their material substance remains intact. The important implication this has for the allocation of resources in a market system is that it creates a basic asymmetry in the way the market functions. While most extractive, harvesting, processing, and distributional activities can be conducted relatively efficiently through the medium of exchange of private ownership rights, the residual mass returned to the environment is almost entirely returned to common property resources, that is, the atmosphere and the biosphere. The same is true of residual energy.

By common property resources are meant those natural assets that cannot, or can only imperfectly, be reduced to private ownership. Examples are the air mantle, watercourses, complex ecological systems, large landscapes, and the electromagnetic spectrum. When open and unpriced (or otherwise unconstrained) access to such resources is permitted, experience indicates clearly what happens. From careful study of specific common property or common pool problems, such as those involving oil pools and ocean fisheries, it is well known that unhindered access to such resources leads to overuse, misuse, quality deterioration, and ultimately, destruction.

Conservation of mass and energy has always been a fact of the natural world. But at lower levels of population size and economic activity, the return of "used" materials and energy to the environment had only localized effects, most of which either did not impinge on other environmental users or could be dealt with by means of ordinances and other local

[4] A residual is a nonproduct (material or energy) output, the value of which is less than the costs of collecting, processing, and transporting it for use. Thus, the definition is time dependent, that is, it is a function of (1) the level of technology in the society at the point in time and (2) the relative costs of alternative inputs at that point in time. For example, manure in the United States is now a residual, whereas thirty or so years ago it was a valuable raw material.

[5] The implications of material and energy conservation for general economic theory are discussed in Robert U. Ayres and Allen V. Kneese, "Production, Consumption, and Externalities," *American Economic Review* vol. 59, no. 3 (June 1969). When the term "consumption" is used here, it refers to a type of activity, not to the dictionary connotation of "being used up," which—as noted—is not physically possible.

governmental measures, for example, those designed to improve sanitation in and around the immediate vicinity of cities. Thus the butchers could be moved, sewers could be installed, and the streets cleared of trash and offal.

But as industrial development and related raw material extraction and processing proceed, more and more material and energy tend to be returned to the environment, and the fourth structural characteristic of a normative competitive economy indicated previously is increasingly violated. Larger "problemsheds" are affected and greater numbers of people more remotely located in both space and time suffer adverse impacts. Common property resources that cannot enter into market exchange suffer progressive deterioration because their use as "dumps" appears costless to the industries, mines, farms, municipalities, and individuals using them that way. This is true even though important values from other uses of the asset are adversely affected or destroyed when the residuals are discharged to them. For example, the destruction of trout fisheries by discharge of residuals to streams, or increased mortality and morbidity from respiratory disease resulting from lowered air quality, nowhere enter into the profit and loss calculations of economic enterprises that obtain their signals and incentives from market prices. In economic terminology, individual economic activities have no reason to consider the "external" costs of production associated with the use of common property resources. The same holds for governmental activities (municipal incinerators) and for individuals (householders, automobile drivers). The externalities resulting from these activities are a primary cause of market failure.

Naturally enough since the "commons" are free, although valuable, no individual user has an incentive to husband them and protect them, and their increasingly deteriorated state yields a narrower and narrower range of services. Ultimately, a watercourse, for instance, may be capable of yielding no service except the carriage of residuals. When such things happen in a market economy, a fundamental failure has occurred in the incentive system.

It is not only that common property resources such as air and water become overused and misused, but opportunities to improve their yield of services are not grasped either. The ability of streams to receive liquid residuals without damage to other uses can often be increased by augmenting low flows through releases of water from surface or groundwater reservoirs and by mechanically introducing air or oxygen into them. A substantial body of research in water quality management makes this clear.[6] In addition, altered ecosystems can sometimes be improved—through introduction of desirable species, for example. Market exchange does not provide incentives for such activities either, and if they occur it must be through collective decisions, investment, and management. In a dynamic context, market exchange does not provide appropriate incentives for technological improvement with respect to the conservation and collective management of common property resources. Few, if any, resources are spent to improve the technological ability to conserve on the use of open-access resources.

In summary, a great asymmetry has developed in the effectiveness and efficiency of existing economic incentives in a market economy. On the one hand, the system works well to stimulate the exploitation of basic resources and to process and distribute them, but on the other hand, it fails almost completely with respect to the management of residuals, especially their disposal to common property resources. The result is that the common property resources are overused, misused, and suffer from deterioration, underinvestment, lack of management, and technological lag. An immediate and important corollary is that there is too much residual material and energy generated and disposed of to the environment; and too little recycling, materials and energy recovery, by-product production, and process and product design to save materials and energy. Consequently, there is too rapid a rate of exploitation of basic resources.

These effects are aggravated by government policies, such as depletion allowances, capital gains tax procedures, and labeling requirements, that discriminate against recovered materials and in favor of basic resource exploitation. All this is in addition to large public subsidies for population increase.

Another result of looking at residuals problems from the perspective of mass and energy conservation, and one that plays an important role in the analysis in this book, is that it helps us understand the nonmarket relationships among different residuals streams. For example, residuals modification processes, such as sewage treatment, clearly do not destroy anything; rather, they merely transform residuals from one form to other types of the same form and to other forms, and change the time and locus of their discharge, or a combination of these processes. In fact, such processes require additional material and energy inputs so that the total mass and energy residuals are increased by their application. Efforts at

---

[6] See Allen V. Kneese and Blair T. Bower, *Managing Water Quality: Economics, Technology, Institutions* (Baltimore, Johns Hopkins University Press for Resources for the Future, 1968).

mitigating problems in one common property resource often contribute to generating problems in another. Management decisions should be made with a full understanding of these interrelationships and in a context where tradeoffs can be explicitly considered and evaluated.

Resource Allocation Theory and Residuals Management

With respect to the residuals aspect of environmental quality problems, a central source of market failure is, as we have just explained, that property rights do not function in the way postulated by the market model. Who owns the air mantle, watercourses, large ecological systems, and large landscapes? Up to now, property rights involving such resources have not been defined or have been defined only vaguely. Accordingly, they belong to everyone and therefore to no one. They are, leaving existing government regulations aside for the moment, open-access, common property resources. Who gets to use such common property resources is not settled by voluntary exchange, as it could be in a market, but by who happens to have physically superior access. The natural world is such that these will be the residuals dischargers. The discharge of toxic materials, for instance, can limit or foreclose the fisherman's use of a stream for fishing, but fishing cannot physically limit its use for residuals discharge. The market provides no way of deciding, as it does with respect to traded commodities or services, whether a more valuable use is being foreclosed by a less valuable one. In the terminology of economic allocation theory, an uncompensated external or "social" cost is being imposed on other users of the common property resource.

The question naturally arises—could not this miscarriage be corrected by defining rights to the common property resource and thereby bringing it into the market system? The answer is "sometimes yes but generally no." Assume that a right to use of the stream for residuals discharge is awarded to the discharging party and that in the case at issue there is only one fisherman who wishes to use the stream. In this instance the fisherman could purchase all or part of that right in the same way he buys any other good or service, and voluntary exchange could produce a "Pareto-optimal" situation.[7] No doubt some situations that have the potential for

[7] A Pareto-optimal situation is one in which all possible gains from voluntary exchange have been exhausted, that is, no individual can improve his own situation without damaging the situation of another, for any given distribution of income. Such an optimum is the outcome of all voluntary exchanges taking place in an ideal market system, as described in the text.

distorting the allocation of a resource are adequately handled by the market in just this way.

But, unfortunately, most of the important environmentally related external cost situations are not of this two-party type. When more than two parties are involved, major complications arise in trying to define property rights and then letting market exchange allocate environmental resources. The reason is that environmental service flows generally have the attributes of what is called a "public good" in the economics literature. That is, they simultaneously yield their service flow to many individuals, in the way that the atmosphere does for respiration. When the quality of the resource is reduced, this service flow is simultaneously reduced for many (perhaps hundreds of thousands or more) individuals. Conversely, when quality is improved, the service flow is increased for masses of people. Furthermore, the use of the service flow by one individual (say for breathing) does not diminish it for others who are using the service flow for the same purpose at the same time. This is the basic distinguishing feature of public goods and is known as "jointness in supply."

Thus a typical situation is as follows: many persons experience the external cost simultaneously. For each one individually the damage cost may be small, but in the aggregate it may well greatly outweigh the cost of controlling it to at least some degree. But since for each individual the damage cost is small relative to the control cost, the individual has no personal incentive to reimburse the waste discharger for curbing it. But the matter is worse yet, for the individual has no reason to cooperate with other injured parties to pay the discharger to reduce the discharge. The reason is exactly that a public good is involved. If for any reason the service flow is improved, the individual automatically benefits from it whether paying for it or not. So the incentive for private gain would never induce the person to participate voluntarily in a scheme to compensate the residuals discharger, even though such a person might well be willing to do so if assured that everyone else would. This is known as the "free rider" problem in public goods economics.

Many people might feel that it would be better if the property right were conferred on damaged parties so that the discharger would have to compensate them for the right to use the common property resource. But this does not solve the problem either. When many parties are involved and the damages to each are individually small, the costs of forming voluntary associations for the purpose of extracting payment are often large relative to individual gains achieved. These costs are known as "transactions costs" in public goods economics. When one envisages the number of sources and receptors of external costs associated with all residuals streams in, say, a large metropolitan area and the complex interactions among them, it becomes clear that private exchange of property rights, even though it has been much discussed in the economics literature,[8] provides no hope of leading toward a Pareto-optimal solution.

To achieve an improved allocation of resources, and to deal with the problem relatively efficiently, we must appeal to collective, compulsory decision making through the medium of government and the execution of management programs whose areal foci have some geographical correspondence with the areas of the various "problemsheds" involved. Consideration of the form and functioning of governmental agencies responsible for such programs will occupy us a great deal in later parts of the book.

One other point merits mention at this time in relation to the discussion of market failure. Let us consider whether it might be possible to identify a set of administered prices for the use of common property resources for residuals discharge that might have desirable properties. It can, in fact, be proven, by using a mathematical general equilibrium model, that there exists a set of such prices that in conjunction with voluntary exchange of private goods will yield a Pareto optimum. Appendix 1-A at the end of this chapter has a sketch of the model Mäler uses.[9] At these prices the incremental external costs (increase in total external costs caused by an additional unit of discharge of each residual) are equated with the incremental costs of reducing residuals discharges (increase in total discharge reduction costs associated with an additional unit of discharge reduction). If there were a management agency and it were able to determine these prices and impose them on residuals dischargers and if it were able to cancel any redistributive effects on income through a system of taxes and subsidies, one could be sure that the result would be a Pareto optimum at a higher level of economic welfare than before intervention occurred.

It is useful to know that in principle there is such a solution to the resources allocation problem associated with common property resources. But one must not be misled by its apparent simplicity. Actually, to approach such a solution in reality would require enormous amounts of

---

[8] A good source on the current state of this discussion is the "Coase Theorem Symposium," *Natural Resources Journal* vol. 13, no. 4 (October 1973).

[9] See Karl-Göran Mäler, *Environmental Economics: A Theoretical Inquiry* (Baltimore, Johns Hopkins University Press for Resources for the Future) 1974.

knowledge about consumer preferences and about how residuals are generated, discharged to, and transported and transformed in the environment. Also recognizing, as explained earlier, that environmental media can themselves sometimes be modified through investment in such a way as to increase assimilative capacity, the technology and economics of all opportunities for doing this would have to be known. Furthermore, conceptual as well as empirical difficulties in evaluating public goods often require that the political process consider economic and environmental quality tradeoffs directly. These are all matters we will consider in detail later.

Some Definitions

The foregoing is a short description of the general nature of the residuals problem and its relationship to the economic workings of society. To complete the introduction and to provide a common basis for the discussions in the subsequent chapters, some definitions are necessary.

Residuals are the "nonproduct" (nonuseful) outputs from human activities. No classification system (taxonomy) of types of residuals will be suitable for all purposes, nor acceptable to all users. Residuals can be characterized by whether they are material or energy residuals, by their state of matter with respect to the former, by the medium in which they are transported, the environmental medium into which they are discharged, or both, and by the type of source and the time pattern of generation. Thus there are: *material residuals*—liquid, gaseous, solid; *energy residuals*—heat, noise; *radioactive residuals*—those having characteristics of both material and energy residuals. The same material, ash or pesticides for example, may be transported in a water stream or in an airstream. Likewise, residual heat may be transported in a water stream or in an airstream. Hence, there are airborne residuals and waterborne residuals.[10] Residuals may be described in relation to the environmental media into which they are discharged or onto which they are deposited, thus: atmosphere; water bodies such as streams, lakes, estuaries, oceans, groundwater aquifers; or land. Residuals may also be described in relation

[10] Generally, in this book the terms "gaseous residual" and "airborne residual," are used interchangeably, as are "liquid residual" and "waterborne residual." But it should be emphasized that residual materials may be transported in fluid streams in dissolved, colloidal, or suspended states.

	MATERIAL			ENERGY		
LOCUS OF GENERATION TIME PATTERN OF GENERATION	POINT	NONPOINT		POINT	NONPOINT	
		Line	Dispersed		Line	Dispersed
CONTINUOUS						
DISCONTINUOUS Regular						
DISCONTINUOUS Irregular						

Figure 1. Taxonomy of residuals

to the extent to which they degrade in natural environments: essentially degradable, that is, organic material; and essentially nondegradable, such as ash, lead, and plastic containers. Many materials do degrade, but at relatively slow rates, such as some long-lived pesticides and some radio-active residuals. The "products" of the degradation may be harmless— such as carbon dioxide and water in the case of organic matter—or worse than the original or primary residual, for example, paraoxon from parathion. Finally, residuals can be characterized by the locus and time pattern of generation. Figure 1 summarizes some of these criteria.

In figure 1, the dividing lines among the classes are not rigid. In some cases the time frame and the geographical area will determine the box into which a residual will be placed. For example, storm runoff from an urban area is generally considered a dispersed, nonpoint residual. However, after the storm runoff has been collected and transported in a pipe, the outlet of the pipe into a stream is in effect a point source, with respect to the environmental medium. The noise from an essentially continuous flow of traffic along a freeway *during the rush hour period* represents a continuous line residual. But if the time period were extended to a 24-hour period, the resulting noise would represent an irregular discontinuous line residual. The dispersed, nonpoint category is probably least applicable to energy residuals. However, the heat discharged in the discharge from

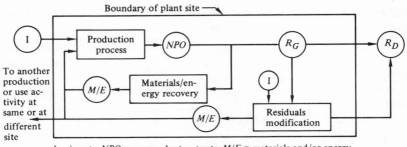

I = inputs; *NPO* = nonproduct outputs; *M/E* = materials and/or energy

Figure 2. Residuals generation and discharge

wild flooding irrigation would fit the characteristics implied by the class.

One additional definitional problem merits attention. Figure 2 indicates the referents to the term "residuals generation" (R_G) and "residuals discharge" (R_D). (I) means inputs. Note that the nonproduct outputs (*NPO*), which—in the absence of what are commonly called pollution controls—are recovered and recycled into the production process, into by-product production, or both, are not defined as residuals. Note also that the discharge of a residual does not occur until the residual is discharged into the environment across the boundary of a plant site, that is, in or from a pipe, stack, truck, or rail car. The residual generated is that nonproduct output produced by the set of processes and operations within the plant. The residual may or may not be modified before discharge into the environment—or transport from the plant site in a sewer line or by truck.[11] A farm, mine, residence (house plus lot), office building, or golf course is each analogous to the plant site.

Keeping in mind the various types and characteristics of residuals, it is logical to consider next the contexts in which residuals problems occur, where context is related to space, time, and political and economic linkages. The first two dimensions are reflected in figure 3. At one end of the continuum there are problems in which the change in ambient environmental quality occurs very rapidly, affecting only a small area within a single political jurisdiction, and where there are no economic linkages

[11] The terms emission and immission are often used, particularly in Europe, the former unfortunately to refer to residuals generation as defined herein, rather than to nonproduct materials and energy "emitted" from (generated in) the production process. Immission, in the sense of injection into the environment, corresponds to residuals *discharge,* as used here.

Figure 3. Time and spatial contexts for residuals–environmental quality management

Geographical extent ↓ \ Time frame →	Short-run	Intermediate	Long-run
Local (micro)			
Regional			
Multiregional			
Continental	?		
Hemispheric (e.g., Pacific Basin)	?		
Global			

Bases for classification

Geographical extent: Spatial extent of impact on environmental quality, i.e., discharge into small tributary stream not affecting the main stem, junked vehicles along main highways vs. on back country lots (mobile vs. nonmobile environmental users), long distance transport in water, e.g., oceans, and in atmosphere (e.g., DDT in Antarctic ice).

Time frame: Length of time for impact on environmental quality to occur, i.e., spread of oil in groundwater, build-up of nitrates in groundwater, change in concentration of CO_2 in atmosphere; essentially the time rate of change of ambient environmental quality; short-run $\frac{dQ}{dt} = \frac{\infty}{1}$, e.g., acid spill into very turbulent stream (very rapid mixing); long-run $\frac{dQ}{dt} = \frac{1}{\infty}$, e.g., oil spreading in slowly permeable groundwater aquifer (slow dispersion).

15

to other areas. At the other end, the residuals problems involve slow, long-run, persistent changes in ambient environmental quality over the entire globe, affecting multiple political jurisdictions and economic activities.[12] Of course, not all situations fit the continuum unambiguously. Some local residuals problems are long-run (persistent), such as the spread of oil from a pipeline leak into a local groundwater aquifer. And a relatively local—in a geographic sense—residuals problem may involve multiple intranational, and even international, political jurisdictions.

By regional is meant other than global. It is necessary to use a word like regional rather than terms pertaining to political jurisdictions, such as nations, states, or cities, because the extent of ambient environmental quality changes resulting from the discharge of materials and energy follow the meteorological, hydrological, and biological systems rather than the boundaries of political systems. The needs for, and ways of, developing regional approaches to residuals management problems constitute one of the main themes of this book.

Appendix 1-A   A Simple Materials and Energy Balance, General Equilibrium Framework

To gain a better understanding of how a decentralized market-type economy works or should work to produce Pareto efficiency, a simple model is used which takes the circular flow of materials into account. The idea is to try to follow the flow of raw materials from the exploitation of deposits in the environment via production processes and consumption processes and back to the environment in the form of residuals. This model is set out graphically in figure 1-A-1. A similar model could also be developed for energy flows.

In the diagram, five boxes corresponding to production, capital accumulation, consumption, environmental protection, and the environment are shown. Before discussing details, something must be said about the environmental protection agency which is referred to later. It is assumed

---

[12] The extent to which the change in ambient environmental quality is irreversible is another important element. For a perceptive discussion of this and other factors used to classify environmental problems, see Clifford S. Russell and Hans H. Landsberg, "International Environmental Problems—a Taxonomy," *Science* vol. 172 (June 25, 1971) pp. 1307–14.

AUTHORS' NOTE: This appendix is based on material prepared by Karl-Göran Mäler of the School of Economics in Stockholm, Sweden, 1972.

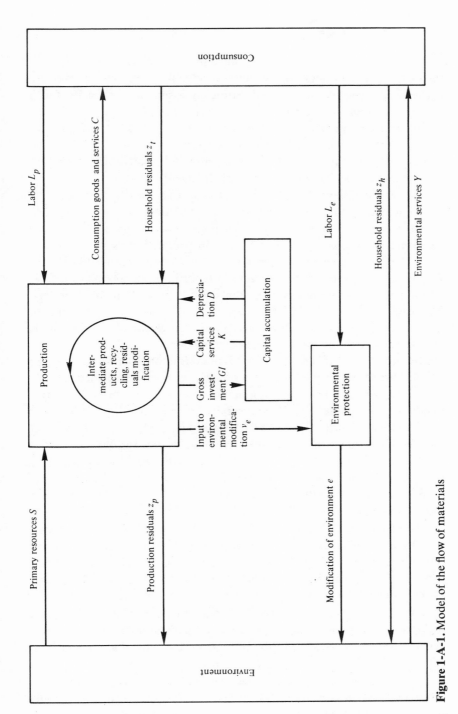

Figure 1-A-1. Model of the flow of materials

that there is an agency whose sole objective is to provide protection for the environment. It does this by charging a fee for using the environment for residuals disposal and by increasing its assimilative capacity (such as by reaeration of water bodies and low-flow augmentation). It is assumed that this modification of the environment is paid for by the consumers and that the total profit of the agency is transferred to the consumers as a lump sum. (The profit may be negative, in which case the consumers pay the expenditures of the agency.)

The diagram shows the flow of materials (and in principle, energy) in the model. In the northwest corner is an arrow S, which corresponds to the extraction of natural or primary nonrenewable resources from the environment. These resources are used in production as raw material. An arrow from the consumption box in the northeast corner to the production box corresponds to the input of labor in production. The capital accumulation box corresponds to the stock of capital in the economy. This capital stock yields productive services that are used as a factor of production. This flow of capital services is represented by the arrow K from the accumulation box to the production box.

Consumption in this model consists of consumption of services, either bought directly from the producers of such services or provided directly by the service flows from consumption goods. The consumers thus buy a bundle of consumption goods and services C, consume the services that this bundle may yield, and are left with residuals equal in weight to what they bought in the form of consumption goods. These residuals are either discharged directly into the environment, represented by the arrow z_h from the consumption box to the environment box, or transferred to the production box where the residuals are modified and raw materials and energy may be recovered.

The capital stock, represented by the capital accumulation box, grows because of gross investments GI, that is, output from the production box set aside for capital accumulation. Because of physical wear and tear, the capital stock will depreciate, however, and there is a flow of residuals D generated by this depreciation from the capital accumulation box to the production box. This flow D has two economic effects: first, it means a decrease in the capital stock, and second, it adds to the flow of residuals generated in the economy. In the production box, raw materials may be extracted from D, and the flow may be recycled.

The environmental protection agency buys labor services and goods represented by the arrows L_e and v_e, respectively, and uses these inputs to modify the environment. The scale of the modification (defined in some way, for example, the rate of reaeration of a water body) is given by the flow e, represented by the arrow from the environmental protection box to the environment box. Residuals modification facilities (sewage treatment plants, incinerators, and the like) are built and operated by businesses and local governments.

Let us now return to the production box. Residuals generated in the consumption box and in the capital accumulation box enter this box. Residuals are also generated in the production processes themselves. The total flow of residuals is modified and partly recycled. This recycling, together with the circular flow of intermediary products, is represented by the circle inside the production box. It is not, in general, technologically possible, and it is still less generally economic, to recover all the residuals, and so there will be a flow of residuals z_p from the production box discharged into the environment.

The quality of the environment is affected by the discharge of residuals and by environmental modification. The environment is regarded as a common property asset, which, in addition to inputs to producers, yields a flow of environmental services to the customers (represented by the arrow Y at the bottom of the diagram). This flow consists of such things as recreation possibilties, aesthetic satisfaction, and clean air to breathe. An increase in the discharge of residuals will in general decrease this flow, while an increase in the extent of environmental protection activities (brought about by reduced residuals discharge or environmental modification) will in general increase the flow.

Let us now adopt the convention that all flows of materials are measured in tons. Using this convention, we can let the symbol C stand both for the flow of consumption goods and services and the weight of this flow. This ambiguity will not create any difficulties, because the meaning will be clear from the context. The same idea applies of course to all other flows as well.

In the production box, the consumption box, and the environmental protection box, no stocks of goods are accumulated in this model, so the arrows entering these boxes must be balanced by the arrows leaving them. We thus have

$$S + z_t + D = C + GI + v_e + z_p \quad \text{for the production box}$$

$$C = z_t + z_h \qquad \text{for the consumption box}$$

$$v_e = e \qquad \text{for the environmental protection box}$$

Adding these identities yields

$$S + D = GI + z_p + z_h + e \quad \text{or} \quad S - (z_p + z_h + e) = NI$$

where NI stands for net investment, $GI - D$

The net outflow of materials from the environment is thus equal to the net accumulation of the capital stock.

This apparently trivial conclusion is, however, very important because it gives the materials balance in a compact form. That part of the raw materials which is not accumulated in the capital stock will ultimately be returned to the environment. Moreover, another conclusion may be drawn. Given the net accumulation of capital, if there are no incentives to keep the discharges of residuals into the environment low, the extraction of natural resources will be too high. If environmental quality is neglected in public policy, then not only will the economy suffer from environmental deterioration, but also from excessive use of natural resources.

To derive more theorems, it is necessary to introduce some kind of mechanism to allocate resources among different uses. So far the model has only consisted of a set of accounting identities, which is useful as a frame of reference but useless when it comes to explaining the working of the economy.

In almost all economies, centralized or decentralized, prices have been the prime regulator of the allocation of resources. Let us therefore introduce prices, some of which are determined by market forces and others by the environmental protection agency, into the model.

It will be assumed that all producers and all consumers are price takers and that the prices for all goods and services, except those for environmental quality and ultimate disposal in the environment, are determined by market forces. The prices for ultimate disposal are determined by the environmental protection agency. The prices for environmental quality are thought of, in this model, as consumers' demand prices for environmental services. Since environmental quality is a public good (as was discussed in the text of this chapter), there are no markets for environmental services, but it is assumed for present purposes that the environmental protection agency knows some method to discover the consumers' demand prices (that is, their marginal willingness to pay for these services).

The following symbols will be used:

p = price of consumption goods and services

p_r = price of natural resources

q_t = price for disposing of household residuals to modification plants

p_K = price of capital goods

p_D = price for disposing of capital residuals by recycling, and the like

q = price for ultimate disposal of residuals in the environment

p_e = price of goods used as input in environmental treatment

w = wage rate

π = capital rental rate

ve = value of environmental modification, that is, the cost of enhancing the environment by a unit

δ = demand price for environmental services

Π_p = profit in the production sector

Π_K = profit in the capital accumulation sector

Π_E = profit in the environmental protection sector

$z = z_p + z_n$, the total amount of residuals discharged in the environment

NS = consumers' net saving

With this notation, we can now calculate the profits in the three sectors (note that the q-prices are prices that the different economic subjects have to pay for the disposal of their residuals).

$$\Pi_p = pC + p_K GI + q_t z_t + q_D D + p_e v_e - wL - p_r S - \pi K - q z_p$$

$$\Pi_K = \pi K - p_K D - q_D D$$

$$\Pi_E = qz + ve - p_e v - wL_e$$

The budget constraint for the consumers is

$$NS + PC + q_t z_t + q z_h = w(L_e + L_p) + P_r S + \Pi_p + \Pi_K + \Pi_E,$$

where it is assumed that all profits are transferred to the consumers. Consolidating these four expressions gives the usual saving-investment identity

$$NS = p_K GI - p_K D$$

Assume now that the relationship among residuals discharges, environmental modification, and environmental services is given by the following environmental function

$$Y = Y_0 - F(z,e) \qquad F(0,0) = 0,$$

where Y_0 is the environmental quality when no residuals are discharged into the environment and no modification of the environment is undertaken. We have already assumed that

$$\frac{\partial F}{\partial z} = F_1 > 0, \qquad \frac{\partial F}{\partial e} = F_2 < 0.$$

The total value of environmental services is δY and the value of the disruption of the environment due to residuals discharge is $\delta(Y_0 - Y)$. This is one factor in the social cost of using the environment for disposal. The other factor is the value of environmental modification ve. The total opportunity cost of using the environment for residuals disposal is thus $ve + \delta(Y_0 - Y)$. The benefits from using the environment for residuals disposal are obviously qz, and the net benefits

$$qz - ve - \delta(Y_0 - Y).$$

Maximizing this net social benefit over z and e yields the optimal policy for the environmental protection agency (it is assumed that the agency regards the prices q, v, and δ as given)

$q = \delta F_1$ and

$v = -\delta F_2$

The agency should thus allow discharge of residuals to such an extent that the social benefit q from dumping one more unit of residuals is equal to the social marginal cost δF_1 of this operation, and the agency should modify the environment to such an extent that the social marginal benefit $-\delta F_2$ equals the social cost v of modification.

For the other sectors it is assumed that there are perfectly competitive markets, in which all decision makers are price takers, who maximize their profits or their utilities. It can be shown that given some additional continuity and convexity assumptions, there exists a set of prices and an allocation of the resources, such that: all markets are cleared and profits and utilities maximized, and the net social benefit from using the environment for residuals disposal is maximized. In other words, there exists an equilibrium for this kind of economy.

Moreover, it can be shown that this equilibrium is a Pareto optimum, in the sense that no person can achieve a better position without leaving at least one other person in a worse position. Finally, it can be shown that all optimal allocations can be represented as equilibrium allocations in this economy. These theorems are proved for the kind of model economy described in this appendix in Karl-Göran Mäler, *Environmental Economics: A Theoretical Inquiry* (Baltimore, Johns Hopkins University Press for Resources for the Future, 1974).

CHAPTER 2

The RFF Quality of the Environment Program

This book is the final product of an evolutionary process of theoretical and empirical research at Resources for the Future on one aspect of environmental quality problems, namely, that relating to residuals generation and discharge. For more than a decade there has been an active research program at Resources for the Future aimed at understanding and illuminating the nature of environmental pollution problems and at devising policies to deal with them. Many scholars participated in this program, and we will have numerous occasions to draw upon their work in all sections of this book. The authors of this volume were primarily responsible for the overall planning and direction of the program.

Evolution

Research on environmental problems at RFF was at first an outgrowth of an older water resources research program. Study of the conditions of water supply and demand in the United States soon led to the conclusion that in many regions of the United States deterioration of water quality was at least as important a problem as the adequacy of supply. Moreover, it became clear that these two problems were often tightly interconnected. Starting in 1960, a series of theoretical, case, and comparative international studies on water quality problems was launched. This phase of research culminated in 1968 with a summary and interpretive RFF volume by the present authors, *Managing Water Quality: Economics, Technology, Institutions.*

Prior to the completion of this set of studies, the first statement of a broader research program was developed by Allen Kneese and Orris Herfindahl in *Quality of the Environment: An Economic Approach to Some*

Problems in Using Land, Water, and Air (1965). This provided the basis for a new program of studies at RFF, launched in 1965 as the Quality of the Environment program. The program initially focused on problems associated with airborne, waterborne, and solid residuals. At the outset these different residuals and the different environmental media were treated as separate categories of problems—even though researchable within a similar conceptual framework. It soon became clear, however, that there were many tradeoffs among the airborne, waterborne, and solid residuals streams. For example—and one could cite many examples— when dissolved organic material and suspended solids are removed from a liquid residuals stream in a sewage treatment plant, a semisolid sludge results, thereby causing a solid residuals disposal problem. If the sludge is incinerated, certain gaseous residuals are generated, thereby transferring the original liquid residuals problem to a possible air quality problem. Conversely, scrubbing particulates from a gas stream by aqueous solutions transfers the residuals from air to water. Or, the modification of a production process to reduce the generation of liquid residuals may involve use of additional energy, thereby resulting in the generation of additional gaseous residuals and waste heat. So far as we are aware, the first study taking explicit and reasonably systematic account of such tradeoffs was the 1968 study *Waste Management: Generation and Disposal of Solid, Liquid, and Gaseous Wastes in the New York Region,* sponsored by the Regional Plan Association of New York but largely conceived, planned, and executed by one of the present authors, Blair Bower.

At the same time that this early empirical work was going on, theoretical research was proceeding on a conceptual framework for more sophisticated work in the area. Some of the basic ideas underlying this work were explained in the previous chapter. The most comprehensive report on the theoretical research is found in the RFF volume by Allen V. Kneese, Robert V. Ayres, and Ralph d'Arge, *Economics and the Environment: A Materials Balance Approach* (1970).

Three basic insights resulted from the early phase of the research program. First, if anything has become clear since the "environment" became popular, it is that environmental quality has many dimensions, and there are almost as many definitions of it as there are those attempting to define it. Included under that rubric is everything from urban design—the effects of the juxtaposition of buildings in space—through vector control, restaurant sanitation inspection, housing standards, to preservation of unique scenic and historic areas. The subsector of the overall environmental quality sector with which our research dealt, and consequently, this book deals, is a major and important subsector—that involving the management of the residuals of society, the "leftovers" from human activities of all kinds. Traditionally, this sector has been termed the "environmental pollution" sector. But we prefer the terminology residuals management and residuals–environmental quality management (REQM).

Second, as noted above, there are basic interrelationships among the three types of residuals—liquid, gaseous, and solid. Further, there are similarities between material residuals and energy residuals (such as noise) in terms of conceptualizing the problem to include: the factors influencing the generation of residuals in the first place, the methods for modifying or reducing the residuals after generation, the discharge and subsequent modification of residuals in the environment, the impacts of residuals on receptors, and the various combinations of incentive mechanisms and institutional arrangements for REQM. Failure to consider the interrelationships explicitly can lead to major inefficiencies in developing strategies to improve ambient environment quality.

Third, it is the institutional problems that are the most difficult with respect to achieving changes in environmental quality and developing adequate environmental quality management. We can detail, and have detailed to some extent, the economic aspects of environmental quality problems, the technological aspects, and—although with less accuracy and precision—the relevant ecological questions. But the major stumbling blocks relate to decision-making processes and the institutional frameworks or contexts in which those processes occur. These insights led to the development of a conceptual framework of REQM on which the detailed research program is based.

Conceptual Framework of REQM

Figure 4 depicts material and energy flows in a society, both as inputs to production and consumption and as residuals from these activities. Figure 5 depicts, in generalized form, the components of the system for managing society's residuals—the REQM system. A short description of these figures will be useful before describing the Quality of the Environment research program based on this framework.

In figure 5, box 0 represents final demand, the totality of goods and services desired by society. Thus, this box encompasses the input of social

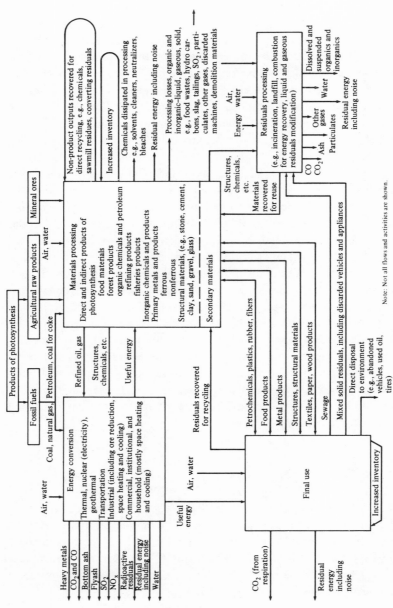

Figure 4. Materials and energy flows in a society. Modified from *Resources for the Future, 1969 Annual Report,* (Washington, D.C., RFF) p. 26.

Note: Not all flows and activities are shown.

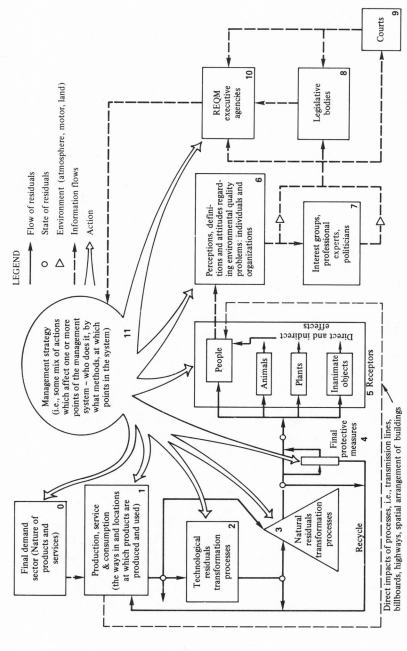

Figure 5. Generalized residuals–environmental quality management system

tastes, mores, habit patterns of society as they are reflected in the multiplicity of goods and services demanded, and the quality specifications for those goods. Final demand is the driving force of the REQM problem.

Box 1 represents the production and consumption activities of the economy, that is, those activities necessary to meet the indicated final demand, and the locations of those activities. (The spatial distribution of activities has very important implications for residuals generation and residuals management.) Included is a vast array of services and energy and materials flows that incorporate the basic extraction of raw materials, their processing into intermediate and final products and distribution to consumers, the transportation of goods and people, and the generation of flows of various residuals. Most of these flows have counterpart market transactions that assign values to them and ration and conserve their use. But the residuals flows from box 2 that interact with common property resources, such as watercourses, the air mantle, and large ecological systems, do not.

Some quantity of residuals flow from a large modern complex economy is inevitable. To reduce the flow to zero would be incredibly expensive. Indeed, it would require an economic system in which all materials are held in closed recycling systems and only solar energy is used. Although some seriously advocate it, with our present knowledge this is an impossible dream. However, this is not at all to say that the amounts and types of residuals flows are fixed. Indeed, even for a given spatial pattern they are highly variable, depending upon: the quantity and quality of final outputs produced, the types of raw materials used, and the production and consumption technologies adopted. All of the variables depend on the technologies available and the costs and prices of particular inputs and outputs. Recognizing this permits us to understand just how deeply the failure of private costs to equal social costs—that is, the presence of externalities—can affect the entire range of decisions pertaining to the production and consumption activities in a market-type economy, and—theoretically at least—in a planned economy. In the absence of any regulation, management, or ownership of the environment, the flows of residuals are limited in size only by the rate of production of goods and services, not by any consideration of the value of the common property resources being used. If restrictions are placed on the use of common property resources for residuals disposal, the incentive structure will change. There will be an added incentive to reduce the flow of residuals into the environment by means of: recycling processes; the use of raw materials with lower, or less harmful, residuals-generating characteristics; and the substitution of products and processes yielding fewer or less harmful forms of residuals.

However, there is the possibility of altering residuals streams further after generation through the complex of procedures labeled technological residuals transformation processes in box 2. It is important to recall from the materials balance principle that residuals modification, as typically practiced, does not reduce the quantity of residuals but only alters their form or location. In fact, as already noted, the total material and energy residuals are increased since the modification processes themselves require inputs.

If the disposal of only certain residuals is restricted, dischargers will respond by transformation processes in which some kinds of residuals are substituted for others. Hence, a policy that aspires to optimality of the whole system must place a price on, or otherwise restrict, all of the residuals streams simultaneously to get the "right" mix of residuals. The discussion of what right means in this connection is postponed until later.

Residuals that are discharged, either directly from the production and consumption processes of society or indirectly from transformation processes of society, enter a natural environmental system. In the absence of collective (governmental) intervention, this is a common property, open access natural resource in a market sense. It may be the atmosphere, watercourses (including groundwater), the land, or the ecological systems associated with any of these. The transport and transformation processes that occur in these natural environments are represented by box 3. These processes convert the time and spatial pattern of residuals discharges into the resulting time and spatial pattern of ambient environmental quality. This box has been labeled "natural" transformation processes to distinguish it from the "technological" transformations which occur in box 2. Models of these natural phenomena have been developed in varying degrees of refinement. These models usually involve the solution of complex systems of simultaneous equations and are used extensively in research and planning activities with respect to REQM. They are central elements in any systematic approach to REQM and are integral parts of the case study of management that is discussed in detail later. Although the present generation of models is useful, there is still substantial room for improvement, particularly at the global scale and also because of the fact that the models were developed for "steady state" conditions. The latter means that variables such as streamflow, temperature, and windspeed and direction are held constant when the model is solved. This, in turn, means that only very limited replications of the highly variable time-dependent behavior of natural systems is possible.[1]

The phenomena in box 4, "final protective measures," include such things as treatment of water prior to use for industrial and municipal purposes, conditioning and filtering of air and the application of sound-proofing in buildings, and application of corrosion-resistant paint and coatings to exposed surfaces. The market system based on individual consumer and business decisions operates rather effectively to implement these measures—clearly more effectively than it does to control residuals discharges. To be sure, an optimal (or efficient) system of REQM would no doubt include many of these measures. However, without effective control of residuals discharges, these final protective measures will have to be induced and used to a larger extent than is optimal or efficient. Visualize, as an extreme example, people scurrying around in gas masks from one sealed and filtered building to another—protective measures —while much less expensive means of controlling particulate discharges go unused.

Box 5 identifies the receptors—that is, those users or potential users of the common property resources who are directly or indirectly injured by residuals discharges to the natural environmental systems and the consequent changes in ambient environmental quality. The extent of injury is a complicated function of many factors: the amounts and types of residuals discharged, the specific transformations in the environment, the extent of use of final protective measures, the type and location of receptors, the physiological condition of individuals as affected by nutrition—and, in the final analysis, the values people attach to the physical, chemical, and biological effects caused by residuals in the environment.

The responses of individuals and organizations to damages are sufficiently complex and important to merit separate identification of the perceptions and attitudes regarding residuals–environmental quality problems (box 6). How an individual perceives a particular level of water quality, along a beach, for example, and his attitude toward expending resources to alter it, are affected by a number of factors. These include

[1] This is not meant to imply that a highly sophisticated model is necessary in all decision-making contexts. The level of sophistication required is a function of the relative importance of all the variables affecting the decision. It should also be emphasized that there is large temporal variability in residuals generation and residuals discharge, as is discussed in chapter 4.

his previous experiences with water quality, his recreational preferences, and his income, among others.

As pointed out earlier, environmental resources have been common property, which means that unless they are somehow converted to private property, management of residuals discharges to them must be executed through some form of collective action. Such action—or nonaction as the case may be—results from the perceptions and attitudes of individuals and groups. Various organizations and institutional mechanisms are possible to initiate the collective action, as is reflected in boxes 7, 8, 9, and 10.

The organizations may be private, such as conservation or environmental associations that bring legal actions, or they may be formal political bodies, such as the U.S. Congress or state and local governments, or the governing board of a river basin agency or air pollution control district. The formal political organizations perform a legislative function. They create public policy, that is, make collective choices. Often the policies devised at the legislative level entail delegation of authority to what are essentially executive agencies, indicated in box 10. Examples include the federal Environmental Protection Agency (EPA), various state departments of environmental protection or environmental conservation, and various regional, subregional, and local agencies that have some degree of responsibility with respect to REQM.

Under principles of strict representative government, these executive agencies are to carry out public policy, not make it. In practice, however, these agencies usually have substantial policy discretion, because most legislation must be made operational by establishing specific guidelines, procedures, regulations, and standards. This is the function of the executive. As a result, policy is in reality created by a combination of legislative institutions and executive agencies. The courts perform a review function, particularly with respect to whether or not governmental agencies have adequately met the procedural prescriptions of the legislation.

The result of these interacting groups is the adoption of a REQM strategy for a particular area or region, or of a residuals management policy for the entire country—represented by box 11. The elements of a strategy are various types of standards—such as secondary treatment at all municipal sewage plants, enforcement procedures, pricing and taxing policies, and private and public investment in residuals–environmental quality management activities. How collective decisions are made and what types of governmental bodies can best make and execute them is one of the most important issues in REQM.

The lines from box 11 to each of the other boxes show the points in the REQM system at which the strategy produces actions that can alter environmental quality directly or indirectly, the latter by influencing the activities of one or more elements of the system. For example, legislation can be passed or regulations adopted barring nonreturnable containers or limiting the amount of phosphates in detergents. In another action, charges can be placed on residuals discharges (box 2) to common property resources, inducing action in boxes 0, 1, and 2: for instance, changing the characteristics of products, using different raw materials, altering production processes, increasing the recycling and transformation of particular residuals streams.

In some instances it is possible to influence directly the external conditions that affect the transport and transformation processes in the environment—hence, the line to box 3. An example is the augmentation of streamflow during low-flow periods and the use of reaeration devices. As the line to box 4 indicates, management agencies may assume a role in providing final protective measures, as when a municipal government constructs and operates a water treatment plant. Management agencies may also have a direct effect on human receptors (boxes 5 and 6), for example, by buying noise rights from people living near airports or by helping receptor groups to organize for action. The effects on plants may result in the agency's selecting or developing species less susceptible to adverse ambient environmental quality (AEQ). The agencies may of course impose constraints on other agencies, changing their institutional and legal structures. Finally, the action of the agencies can involve changes in the agencies themselves (box 10).

The Program

In the context of the preceding framework, the organization of the Quality of the Environment research program can be described in two ways. One is in terms of the three basic thrusts of the program: (1) micro and regional analyses, (2) collective choice theory, and (3) basic economic theory and macroanalyses. The other is in terms of studies designed to elucidate one, several, or all elements of the REQM system with respect to one or more residuals. The two ways overlap, as will be evident in the following description.

Micro and Regional Analyses

Microanalyses in this book refer to studies of residuals generation and management in individual activities, such as an industrial plant. A series of detailed studies of individual industries was undertaken that involved the simultaneous analysis of all residual material and energy flows (in three environmental media) and a wide range of technological alternatives to meet specified sets of residuals control targets. Industries included were petroleum refining, steel, pulp and paper, and fruit and vegetable canning. Two related industry-type studies involved the analysis of energy generation from coal (in coal-fueled power plants) and by liquid metal fast breeder reactors, where residuals generation and management in the total fuel cycle were analyzed, that is, from extraction through generation and handling of spent fuel.

Regional analyses were of two types. By far the most important was the building of a quantitative regional REQM model. This model is intended to be useful to both legislative and executive agencies for the analysis of: alternative policies for residuals management, such as direct regulation versus effluent charges; the relationships among costs, level of environmental quality, and type of management strategy; and the distributions of costs and benefits associated with alternative policies, that is, whose environmental quality is in fact improved by a given policy or strategy[2] and who bears the costs. Furthermore, such a quantitative model permits testing the adequacy of governmental structures designed to manage the residuals components of environmental quality, for example, the performance characteristics of regional or metropolitan specialized agencies of general governments versus more loosely linked councils of independent local governments versus regionalized agencies of state governments. This model was applied to the Lower Delaware Valley region (Pennsylvania, New Jersey, Delaware). A very important part

---

[2] The terms "policy" and "strategy" are widely used and seldom defined. Operationally, the difference is in the specificity and scope of application. Strategy connotes a detailed identification of particular measures to be carried out in a given area; policy connotes a course of action to be applied to broad classes of activities throughout the jurisdiction, for example, federal financing of X percentage of municipal sewage modification plant construction costs.

of this regional study was the development of an aquatic ecosystem model of the Delaware Estuary in the Lower Delaware Valley. A regional study similar in concept, but with very small resource inputs, was made in the Ljubljana region in Yugoslavia.

The second type of regional analysis involved a study of a particular type of residual in a specific region—used newspapers and used corrugated containers in the Washington, D.C., metropolitan area in relation to recycling; obsolete automobile processing to produce steel scrap, in the Philadelphia metropolitan area; dispersed solid residuals in New Mexico.

Collective Choice Processes

Given the crucial role of decision-making processes in REQM, a set of studies was undertaken to provide both a theoretical base and some empirical analyses of collective choice processes. The former included analysis of vote trading in relation to structure and representation of legislative bodies. The latter included analyses of: the role of the courts, the National Environmental Protection Act, and regional power structure in the northeastern megalopolis in relation to REQM decisions. In addition, descriptive studies of institutional mechanisms in Florida, the United Kingdom, and San Francisco Bay were made.

Economic Theory and Macroanalyses

The primary work in economic theory involved broadening such theory to incorporate residuals explicitly, by combining the principles of mass and energy balances from physics with general equilibrium theory from economics. Given the basic theory, several macroeconomic issues were addressed. (Macroeconomic refers to impacts of large-scale changes, in this case those associated with environmental deterioration or environmental policy, on certain aggregate measures of economic conditions or performance.) Under this rubric were included the following studies: how aggregate measures of economic performance, such as gross national product, might be improved in the presence of environmental deterioration and large-scale expenditures for its avoidance; how social indicators might be combined with national accounts; and the basic economic theory of recycling.

The final type of macro study involved use of an input–output (I–O) model of the U.S. economy to analyze the interrelationships among population, resources, the economy, and residuals. Despite the inherent limitations of such a model, the approach is useful because it permits at least some estimates of the indirect (interindustry), as well as direct, effects of major changes in the economy. This methodology was embodied in an aggregative model for projecting major components of national final demand. It was designed to make some projections of future residuals generation and discharge resulting both directly and indirectly from changes in the size and structure of final demand, as final demand is affected by economic and demographic variables (especially growth) and by resources devoted to residuals management.

Other Studies

A variety of other studies was undertaken to shed light on problems of managing particular types of residuals and residuals generators or to illuminate particular issues. Examples of the former include studies of pesticide use in agriculture—productivity, externalities, management options; noise management in urban areas; and alternatives to the internal combustion engine in motor vehicles. Examples of the latter include studies of: effects of gaseous residuals on human health; the distribution of air quality among different socioeconomic groups; and the differential perceptions and attitudes about smoke control programs.

A bibliography of publications resulting from the program appears at the end of the book.

Overview of the Book

Results of many of these related lines of work are discussed in this book. Although each study is not discussed in detail, each has been drawn on directly or indirectly.[3] The presentation is primarily expository and quantitative rather than mathematical, even though much of the underlying work has involved extensive use of mathematical models. For those who are interested in the models and proofs, we provide a highly selected set of technical appendixes, as well as references to more detailed discussions.

[3] That being the case, no attempt is made to identify specifically all of the studies and publications or the contributions of specific individuals, as we describe the results throughout the book.

The book is divided into five parts. Part one is comprised of chapter 1, which explained in terms of our conceptual framework why there is a residuals *problem*, and this chapter, which explains the evolution, framework, and content of the research program. Part two consists of chapters 3 and 4. The former is a general presentation of integrated residuals management at the micro (that is, plant) level and extensions of such management. Included is a taxonomy of measures available to improve environmental quality, both by controlling the generation and discharge of residuals and by improving or making better use of the assimilative capacity of the environment. Chapter 4 is a discussion of technologies and costs of integrated residuals management in particular industries.

Part three focuses on regional analyses and consists of three chapters. Chapter 5 and its appendixes deal with the formal structure of the model developed in the Quality of the Environment program for the analysis of regional systems. Here is the framework for bringing together the economic, technological, ecological, institutional, and political concepts and information developed in the research performed during the program. Chapter 6 and its appendixes focus on a more extended theoretical discussion of the institutional and political aspects of regional REQM. In the same way that a modified theory of economic general equilibrium provides the conceptual structure for analysis of the economic, technological, and ecological aspects of the problem, collective choice theory provides the underpinning for examination of the political and institutional aspects.

The final chapter in part three, chapter 7, presents the results of the research that has been at the center of RFF's Quality of the Environment program for the past several years—the illustrative application of a detailed, quantitative, regional REQM model to an actual region. As already mentioned, the model developed for the application includes: the technological and economic considerations involved in simultaneously controlling waterborne, airborne, and solid residuals streams; the effects of residuals discharges on natural environmental systems; the effects of REQM strategies on ambient environmental quality and distribution of costs of space heating, electricity, and solid and liquid residuals disposal; and political collective choice aspects of selecting levels of environmental quality and institutional forms for residuals–environmental quality management. The case area is a region centered in the Delaware Estuary.

Part four deals with macroanalyses and is comprised of two chapters. The first, chapter 8, provides a general quantitative perspective on the residuals problem in the United States now and in the future and a broad view of the costs that may be involved in dealing with it, utilizing an input–output model of the U.S. economy. Chapter 9 considers whether our national accounting system should be altered to reflect environmental deterioration and the costs of REQM. Part five, which consists of chapter 10, compactly reviews the main findings and insights.

PART TWO

Microscale Issues and Analyses

CHAPTER 3

Integrated Residuals Management at the Microlevel: General Considerations

In chapter 2 we presented a schematic diagram of an overall residuals–environmental quality management (REQM) system. This and the next chapter focus on the meaning of integrated residuals management with respect to box 1 of that diagram—the production and consumption activities in the economy—and with respect to the interrelationships among boxes 0, 1, and 2. We do this by analyzing residuals generation and management at the individual activity level, looking at various options for reducing residuals discharges and their costs. Such analyses are essen-, tial to analyses of REQM at both micro (regional) and macrolevels. This chapter provides a qualitative description of some of the important factors affecting residuals generation in society and the effects of changing those factors; the next chapter describes quantitative studies of individual activities.

In Production and Consumption Activities

The forms (liquid, solid, gaseous) of residuals generated in individual activities can be changed by modification processes as can, to some degree, the types of residuals (mass versus energy). The amounts and composition of residuals can also be affected by the types of raw materials used; the product mix and product specifications;[1] the introduction of materials-

[1] We find it useful to distinguish between product mix and product specifications. For example, a product mix change for a paper mill producing consumer products would be a shift from manufacturing 50 percent paper towels and 50 percent tissues to producing 25 percent paper towels and 75 percent tissues. For any *given* product mix, there could be different specifications for each of the products, that is, GEB 80 or GEB 25 for the paper towels. Similarly, a petroleum refinery might shift the proportions of gasoline and kerosene or jet fuel produced, at the same time maintaining or changing the specific characteristics of the gasoline, or kerosene produced, or both.

TABLE 1. Classification of Physical Methods for Reducing Residuals Discharges (with particular reference to manufacturing activities)

Category and subcategory of physical method	*Examples*
PANEL A. REDUCE RESIDUALS GENERATION	
1. Increase longevity of goods	
2. Change raw materials inputs	High to low sulfur crude, fuel oil, coal; concentrated vs. raw ore; residuals instead of virgin materials, i.e., aluminum cans instead of bauxite, waste paper instead of virgin wood; short-lived, specific pesticides instead of long-lived, general pesticides in agricultural production.
3. Change production process, including mode and motive power of transport	Individual vehicles to mass transit; ICE to ECE; H_2SO_4 to HCl for pickling steel; CEHDED bleaching to oxygen bleaching; ingot casting to continuous casting; less energy-intensive production processes, i.e., for producing aluminum; less water- and energy-intensive appliances in residences; feedlot-fed cattle to range-fed cattle; clear-cutting to selective cutting; minimum tillage instead of intensive tillage agriculture; intermixed cropping (for pest control) instead of uniform cropping; dry peeling instead of wet peeling in canning.
4. Change final demand	*a.* Change product mix—reduce number of grades or styles of product, i.e., chemicals, linerboard, paperboard, canned peaches, models of automobiles.
	b. Change product specifications—reduce brightness of consumer paper products, such as towels, napkins, tissues, and printing/publication papers; reduce phosphate content in detergents; reduce lead in gasoline; high octane to low octane gasoline; returnable instead of nonreturnable beverage containers; change grading to permit a few blemishes on fresh fruits and vegetables; reduce complexity of product, i.e., number of parts in automobile, appliance.
5. Change timing of activity	Stagger office hours; change production schedule.
6. In-plant recirculation of water[a]	In beet sugar production, peach canning, petroleum refining.

(continued)

Table 1. (continued)

Category and subcategory of physical method	Examples

PANEL B. MODIFY RESIDUALS AFTER GENERATION IN ON-SITE AND/OR COLLECTIVE FACILITIES

1. Materials recovery[b] (direct recycle)	Chemical, fiber, heat recovery in paper production; recycling of mill scale in steel production; recovery of used lubricating oil.
2. By-product production[b]	*a.* To final products—tomato pulp into pet food; citrus peels into candy; peach pits into charcoal briquettes; wood products residues into pressed logs; fly ash into bricks; cheese whey into animal food; tree trimmings into mulch.
	b. To intermediates—obsolete vehicles into steel scrap/steel; used corrugated containers into linerboard; used aluminum cans into aluminum ingots; sulfite waste liquor to yeast; used newspapers into newsprint.
3. Modification of residuals streams	Combustion of solid residuals to generate energy; incineration; landfill; composting; compression of solid residuals; land spraying of sludge; precipitation; sedimentation; scrubbing; biological oxidation; chemical oxidation; constructing artificial topography with solid residuals, i.e., hills, reefs with used tires, obsolete auto hulks; spray irrigation; regrading, revegetating strip-mined areas.
4. Effluent reuse	*a.* Direct—sewage plant effluent for cooling water.
	b. Indirect—groundwater recharge with modified liquid residual.

[a] Only in relatively few types of cases does in-plant water recirculation modify or reduce a residual in a liquid residuals stream. Recirculation does reduce the hydraulic load on any materials recovery or residuals modification facility, thereby reducing residuals modifications costs.

[b] In addition to any materials and energy recovery and by-product production that would have taken place in the *absence* of constraints on residuals discharges.

or energy-saving technologies, or both; materials and energy recovery (in-plant reuse); by-product production; and various types of indirect recycling operations.

The methods of reducing residuals discharges from production and consumption (aside from simply reducing these activities) can be grouped into two broad categories, as shown in table 1. Although the examples in the table relate primarily to manufacturing activities, the classification itself is equally relevant to agricultural, mining, forestry, commercial, institutional, and residential activities. Because, comparatively speaking, much has been written about so-called treatment processes after generation, we select a few of the other most salient possibilities for general discussion here. This serves as an introduction to the more detailed consideration of specific industrial activities in the following chapter.

Design of Products and Quality of Inputs Used

As noted in chapter 2, final demand is the driving force of the residuals problem. Hence, the design of products (qualitative characteristics or product specifications) can greatly influence the amount and type of materials used in production and the types and quantities of residuals generated. For example, producing a white paper of high brightness (about General Electric brightness [GEB] 80) requires substantially greater quantities of chemicals, water, and energy and accordingly results in the generation of larger amounts of some residuals than an unbleached paper (about GEB 25) otherwise similar in quality. By using the kraft process to produce tissue paper, liquid residuals of dissolved inorganic solids and dissolved organic solids would be about 90 percent and 85 percent less, respectively, if no bleaching were required.

Similarly, the quality of inputs used can significantly affect amounts of materials disrupted, extracted, and dissipated to the environment. When a high quality natural gas is burned, it comes close to releasing only CO_2 and H_2O as combustion products, and the evidence of its extraction can be limited to some relatively small holes in the ground, plus some landscape disruption from exploration. When coal is used for a similar purpose, vastly greater amounts of material and energy are involved all along the way. If coal is obtained by strip mining, there is overburden from extraction, and there may be residuals from processing the coal; if coal is obtained by deep mining, there may be acid mine drainage and land subsidence, as well as residuals from coal processing. The combustion of coal produces a rich complement of sulfur dioxide (SO_2), oxides of nitrogen (NO_x), particulates, bottom ash, and other unwanted materials —including in some cases heavy metals and radioactive materials. The quantity of SO_2 generated in combustion is a function of the sulfur content of the raw coal and the extent of its removal, if any, in coal processing.

Another specific example is found in petroleum refining. Residuals generation in that industry is sensitive to such factors as the sulfur content of the crude charged and of the specifications of the products produced. For instance, shifting to low-lead, high-octane refining from high-lead, high-octane refining results in about a 15-percent reduction in biochemical oxygen demand (BOD_5) generated per barrel of crude charged. Clearly, one important area of adjustment in materials and energy use and residuals generation is the qualitative character of inputs and outputs.

The design of products with respect to durability and repairability also affects material and energy flows. Repairability seems to be particularly important, and, in a high labor cost economy, modularity seems to be an essential for achieving this end. One can visualize many tasks now carried on by complete integral machines being done by modular—possibly multipurpose—components. These modules would be repairable and replaceable individually. Today even relatively complicated machines, such as television sets, and most other household appliances, are thrown away because of the malfunctioning of one component or a few integral components that are difficult to locate and replace. Because, from a materials and energy point of view, the most efficient form of recycling (to somewhat stretch the word) is that which involves the least change in form or composition, it is apparent that repair or rebuilding of modules would be highly favored over replacement by new or different types. A slight tendency in this direction is already evident. Product specifications also have important implications for recycling—a subject that is treated in the discussion of recycling below.

Process Changes Leading to Increased Recovery

Fuller recovery of primary and by-products (that is, converting more of the inputs into products) is one of the most important activities available for improving efficiency of materials and energy use. Often larger output of the primary final product for a given input can be achieved, and many instances exist where this has already been accomplished, given the proper stimulus. Examples abound in operations involving chemical processes, in which chemicals previously discarded in effluent streams are now recovered for reuse in the same process or in other processes. In the beet sugar industry innovations have resulted in a substantialy higher sugar recovery from the beet. The same is true in the canning of peaches, where one-third more of the raw peach now is included in product output than was the case two decades ago. Most of these innovations came about without the stimulus of residuals discharge restrictions. Many opportunities remain to be taken advantage of if the external costs of production can be brought into the economics of process design decisions.

In addition, materials that are technically incapable of being used more efficiently in the production of the plant's primary products constitute a primary potential source of by-products that may be produced in the same establishment or in others to which they are transferred. United States industrial history presents a large number of examples of successful by-product development. Just a few of these include: the transformation of slaughterhouse residuals into valuable raw materials for the pharmaceutical industry; the development of products under the general heading of "silvi-chemicals" based upon raw materials which originally were contained in wood pulping residuals; the production of animal foods from brewery, distillery, and food-processing industry residuals; the utilization of power plant flyash, used glass, and recovered sulfur in the production of building materials and airport runways; the recovery of textile residuals for use in paper manufacture; and the collection of animal fat from slaughterhouses and food preparation establishments for use in rendering plant operations in the production of soaps, glues, gelatin, glycerin, and a variety of other products. Again, one can expect to see many further adjustments of this type when external costs become internalized.

However, it should be emphasized that the design of many manufacturing processes is based on that *combination* of production process efficiency (physical conversion of inputs to products) and materials recovery efficiency (physical recovery of materials and energy for direct reuse in the production process) which maximizes profits. Because both production processes and materials and energy recovery processes generally have economies of scale, and because the extent of materials and energy recovery commercially justified is affected by relative factor prices (particularly the cost of "new" materials and energy), the extent of recovery is time dependent. Hence, for any given unit process or operation at a point in time, none, some, or much materials and energy recovery for direct recycling will be economically feasible even in the absence of pollution controls.[2]

[2] For a detailed discussion of a specific example, see George O. G. Löf, W. M. Hearon, and Blair T. Bower, "Residuals Management in Pulp and Paper Manufacture," in Walter S. Kaghan, ed., *Forest Products and the Environment,* American Institute of Chemical Engineers (AIChE) Symposium Series, no. 133, vol. 69 (1973).

Recycling Materials That Have Entered into Final Goods

A major form of recycling is the reuse of a material that has been incorporated in a final good. We used the word "final" here in the sense of the national economic accounts (see chapter 9). That is, we do not limit it to consumer goods but refer to goods that are at the end of their processing (and use) sequence. In this sense, both a drill press and an automobile are final goods.

Materials in residuals represent alternative "raw materials" to virgin materials, including the energy content of residuals. The extent to which recycling of the material, as material or energy, back into production processes occurs is a function of the relative prices of the alternative raw materials.[3] As with materials recovery and by-product production, relative prices are time dependent.

Factors influencing the value of a residual as a raw material include: the quantity generated at the source; the homogeneity of the residual; the amount of contamination, that is, other materials associated with the desired material; the cost of processing for reuse; and the location of generation in relation to the locations of use. Thus, used newspapers kept separate from other household solid residuals in an apartment building of 200 dwelling units have much greater potential for being recycled economically to obtain cellulose fiber than used newspapers mixed in with all other solid residuals collected in a municipality. Consideration of separation at the source is an important element in recycling.

As is evident from the above, recycling of materials in final goods typically involves a number of different enterprises at various processing and distribution stages of the production→use→residuals-handling cycle. One of the simplest of such cases is the recovery of lead from discarded automobile batteries, in which the used batteries typically are channeled backwards in part through the same distribution system as the new batteries were channeled forward. There exist organized and, in many instances rather complex, specialized recycling (secondary materials) industries for virtually all of the major metals, including not only lead but scrap iron and steel, copper, aluminum, zinc, silver, and others. In addition to the metals, various types of waste paper, rubber, glass, and plastics are all recycled to some extent through specialized channels of collection and reprocessing. In some instances, as with battery lead, the materials are reused in

[3] These relative prices, it should be emphasized, are affected by various government policies relating to taxes, depletion allowances, and freight rates.

processing plants that also utilize primary virgin materials. In other instances, the secondary products are processed by specialized processing plants dealing only with recovered secondary raw material, and which compete with primary producers of similar or competing raw material.[4] In some cases, particularly containers, the item itself is reused. Examples of this are refillable bottles, reusable pallets, and returnable wood, plastic, and metal shipping containers.

As noted in the previous section, product specifications can also substantially affect the extent to which materials can be recovered for recycling. For instance, a major obstacle to recycling steel from obsolete automobiles is contamination with foreign metals, particularly copper. This problem could be substantially reduced by a variety of measures, for example, making wiring of (larger diameter) steel rather than copper. In addition, the characteristics specified for the steel for automobile bodies themselves exacerbate the problem. To achieve the ever smaller gauge (lighter weight) steel Detroit specifies for automobile bodies requires the addition of more nonferrous materials to achieve the requisite strength. This increases the cost of producing steel from raw scrap, reduces the quantity of scrap that can be used in making certain types of steel, or both.

A similar problem of "contamination" stemming from product design occurs in bimetallic containers, such as those made of steel plus aluminum. Such combinations preclude economical processing to obtain materials for reuse, at least at present. Another aspect of the effect of product specifications on recycling is illustrated by the production of various paper consumer products—towels, napkins, tissues. Requiring high brightness (about GEB 80) in the product means that a smaller proportion of waste paper (paper residuals) can be used as raw material.

Still another aspect of recyclability related to product design is the great diversity of product types that often have very little apparent utility. An illustration of this is the multiplicity of glass container sizes and shapes.[5] As an antidote to this, the National Bureau of Standards might

---

[4] The residuals, that is, waste paper, discarded copper wire, used aluminum containers, are raw materials competing with counterpart virgin raw materials, that is, trees, lead ore, copper ore, and bauxite, respectively. A residual may be processed to obtain the relevant material, that is, cellulose, lead, copper, aluminum, in the plant which produces the final product, that is, paper product, aluminum can, or it may be processed by a secondary materials processor who sells his raw material output to the fabricator. As in all other activities, the processing of residuals to produce useful materials itself results in the generation of material and energy residuals.

[5] The primary stimulus to such diversity has been "product identification."

specify a short list of standard containers that would be interchangeable from brand to brand of product contained in them. There would be a standard 12-ounce beverage bottle (maybe one transparent and one green), several standard shapes of quart jars, and so forth. Under these conditions, the costs of creating a market in recycled containers would be substantially decreased.

In this discussion we have not included the most straightforward reuse process of all, the secondhand goods market. The reason is that we do not, for present purposes, regard materials as residuals as long as they are salable, but secondhand goods require mention at least because of the substantial variety and total volumes of materials involved and their potential variability over time. This form of reuse primarily involves durable goods, both of the household and producers' equipment types, and could also be defined to include houses and other buildings. The technical processes involved are generally limited to some combination of transportation, storage, and possible repair prior to resale, but the principal function is that of brokerage. Except for items subject to the cultural phenomenon of "antiqueness," much of this activity, especially that related to consumer durable goods, represents a "trickling down" process of transfer from the more wealthy to the less wealthy members of the community. The importance of used-product resale to problems of residuals is obviously that of increasing the economic durability of the commodities involved and postponing their time of discard or demolition. The recent great boom in flea markets, garage sales, and basement sales merits mentioning in this connection. Similarly, there is an extensive market for used producer goods, ranging from sawmill equipment and canning equipment, to desk calculators, adding machines, and typewriters, not to mention airliners and military aircraft.

Innovations That Reduce Material and Energy Needed
per Unit of Service Rendered

In recent years there has been a tendency, especially in the electronics field, toward technological innovations that reduce the amount of materials and energy needed to render a given level of service. For example, solid state devices have replaced vacuum tubes and printed circuits have replaced wiring in many applications from small radios to giant computers. A similar statement can be made about the electronic in contrast to the mechanical desk calculator. Electric drives for machinery have replaced bulky and inefficient mechanical energy transfer systems in factories. As the cost of materials rises, both as an input and for residuals disposal, there will be a stronger incentive to move in this direction, but it seems doubtful that we can expect many innovations with the far-reaching implications of the transistor and the electric motor. Nevertheless, such possibilities as laser beam communication and cryogenic transmission of energy are on the horizon and could be greatly stimulated by the internalization of external costs.[6] On a more prosaic level, developments of process technology in the aluminum industry, as one example, have occurred which have resulted in a substantial decrease (20 to 25 percent) in energy used per ton of aluminum produced.

Possibilities are not limited to the electronic and electrical field. Design of products and innovations in materials can have a similar effect in other areas. Tonal quality of a high order can be achieved by a small infinite baffle speaker and matched only by a comparatively huge bass reflex. A geodesic dome encloses a given area with only a fraction of the mass of materials required to enclose a similar space with a conventional building. A much older example is the contrast between the strength of steel and that of iron. A newer one is high-strength, special-purpose aluminum. In the energy field, fuel cells hold promise for much more efficient use of fuel—a very large material flow in our economy.[7] Finally, the mass of many containers could no doubt be reduced by the application of materials technology.

Concluding Comments

We have only skimmed the surface in this general discussion of materials- and energy-saving and residuals-reducing products and technologies. We know the range of these alternatives is large because actual examples of each type abound. Two additional points are relevant to the discussion of the conceptual structure for analysis.

[6] Sometimes, miniaturization makes use of very rare materials—gold in electrical circuits, for instance. In these instances, large amounts of materials may have to be processed to get a very small amount of the desired one.

[7] Mineral fuels constitute more than half of the weight of basic materials production in the United States. A summary of these weights can be found in Allen V. Kneese, Robert U. Ayres, and Ralph C. d'Arge, *Economics and the Environment: A Materials Balance Approach* (Baltimore, Johns Hopkins University Press for Resources for the Future, 1970).

First, the net environmental effects of any of the alternatives discussed above—or any others—can be adequately assessed only in the context of the total system. Such a system must include the residuals associated with all of the material and energy inputs relating to the production of the good or service and all of the material and energy inputs, and related residuals, necessary to handle or dispose of the good subsequent to the end of its useful service. Thus, whether or not increasing the longevity of products will result in positive net environmental gains depends on the inputs required and the resulting environmental effects "all along the line." Rigorous analysis of such total systems would involve not only individual activity models but also some type of interregional "product" model, such as that developed for certain agricultural crops by Heady et al.[8]

Second, the preceding discussion has focused on reducing discharges as the mechanism for improving ambient environmental quality—quality which can also be improved by undertaking measures that directly affect the assimilative capacity, rather than changing the quantity of residuals discharged per se. The following are examples:

Measures for making better use of assimilative capacity
1. Effluent redistribution over space, over time, or both, construction of higher stacks
2. Change in the time scheduling of activities (where unit generation does not change)
3. Change in the spatial location of activities, for example, land use zoning
4. Construction of artificial topography with solid residuals

Measures for increasing the assimilative capacity
1. Addition of dilution water
2. Use of multiple outlets from reservoirs
3. Artificial mixing of water in reservoirs and lakes
4. Artificial aeration of streams, lakes, and estuaries with air or oxygen, by surface or subsurface diffusers

[8] Earl O. Heady, Howard C. Masden, Kenneth J. Nicol, and Stanley H. Hargrove, *Agricultural Water Policies and the Environment: An Analysis of National Alternatives in Natural Resource Use, Food Supply Capacity, and Environmental Quality* (Ames, Iowa, Center for Agricultural and Rural Development, Iowa State University, June 1972) (Card Report 40).

Measures for reducing residuals discharges and measures for affecting assimilative capacity directly are both relevant to residuals–environmental quality management. The ways in which they are integrated in a regional analysis are illustrated in the later discussion of the Delaware case study.

Although this wide range of alternative products and technological options is available, such availability does not indicate that either the use of those known options or the development of new options is adequate compared to what both use and development would be if the external costs of using common property resources for residuals discharges were effectively made internal to the decision-making processes of private and public units. A major problem for public policy is to develop incentives and institutional mechanisms to internalize these costs in such a way that the most efficient combinations of these products and technologies will in fact be selected and applied at the appropriate levels.

Chapter 4 focuses on integrated residuals management at the plant (micro) level in several industries that have been studied in detail in the Quality of the Environment program. These should be taken as illustrative of the possibilities for adjustment in the whole range of human activities affecting the environment. A complete analysis would certainly have to include mining, agricultural (including forestry), household, and transportation activities. But the industries which are discussed reveal the richness of the adjustment possibilities and, by implication, the weakness of policies that arbitrarily focus on one or a few possibilities.

CHAPTER 4

Integrated Residuals Management in Industry

We now turn to a discussion of specific case studies of industries. Because each of the industries studied represents a degree of complexity that cannot be completely reflected here, we present only a sampling of approaches and results, emphasizing the utility of such studies and what has been, and can be, learned from them. Because the studies considered here are the result of a more or less continuous evolution in concepts and methodology in the RFF Quality of the Environment program over a period of some ten years, it will be worthwhile to recount a little of this history.

Introduction

The first efforts at detailed industry studies focused on the demand for water as an input to production processes. They reflected, first, the widespread concern at the time with the possible scarcity of resource inputs to support continued economic growth, and second, the fact that previous efforts with respect to estimating future industrial water needs had been done on a very rudimentary and naive basis. Traditionally, industrial water needs had been estimated on the basis of historical information on gallons per employee, gallons per unit of raw product processed, gallons per unit of final product output, and even gallons per acre of type of industrial activity, for example, "light manufacturing." The basic data used were aggregate data, that is, across a given industry and based on nationwide mail questionnaire surveys, such as those of the Bureau of the Census. In general, none of the estimates by public agencies and private firms considered explicitly the effects of: (1) changes in production technology, raw materials, product mix, and product output specifications; (2) the price of water at both intake and outlet, as price affected industrial water utilization through the many substitution possibilities available in industrial water utilization systems; (3) the prices of other factor inputs, such as electrical energy, fuel, and raw materials other than water; and (4) various factors external or exogenous to the individual plant, such as various tax policies and grade labeling of products. For these reasons, RFF undertook research to develop more rational bases for estimating future industrial water demands. The Cootner-Löf study of the steampower industry[1] is the major example of published work from this first period of research.

The next set of industry studies continued the concern with water, but the focus was broadened to include not only questions of water intake and in-plant water utilization, but also liquid residuals generation, modification, disposal, and discharge. This reflected the emerging recognition that a major, if not the most important, impact of industrial water use on water resources is the degradation of quality caused by residuals discharges. The liquid residuals and water quality parameters of concern were biochemical oxygen demand (BOD_5) and suspended solids (SS). A study of the beet sugar industry by Löf and Kneese[2] is the primary product of this period of research. Even though the beet sugar industry is a relatively simple one, much was learned in the study about methodology, and the study itself generated information that has been widely used.

A study of residuals management in the New York region for the Regional Plan Association,[3] in which the RFF Quality of the Environment program was heavily involved, provided the stimulus for a further expansion of the framework of the RFF industry studies. This New York study made clear the potential importance of the basic technological, physical, and economic interrelationships among the two basic types of residuals—materials and energy—and the three states of the former, liquid, gaseous, and solid.

The result was that subsequent RFF industry studies focused on the simultaneous management of all residuals generated, as well as water

[1] Paul H. Cootner and George O. G. Löf, *Water Demand for Steam Electric Generation: An Economic Projection Model* (Baltimore, Johns Hopkins University Press for Resources for the Future, 1965).

[2] George O. G. Löf and Allen V. Kneese, *The Economics of Water Utilization in the Beet Sugar Industry* (Baltimore, Johns Hopkins University Press for Resources for the Future, 1968).

[3] Blair T. Bower, G. P. Larson, A. Michaels, and W. M. Phillips, *Waste Management: Generation and Disposal of Solid, Liquid and Gaseous Wastes in the New York Region, a Report of the Second Regional Plan* (New York, Regional Plan Association, 1968).

utilization, in the industrial plant. Studies of petroleum refining,[4] steel manufacture,[5] pulp and paper manufacture,[6] steel scrap production,[7] and the coal-electric energy industry[8] are examples of this expanded research focus.

The main objectives of the industry studies, as they have evolved during the progress of the studies themselves, are as follows:

1. to identify both the endogenous and exogenous factors that influence residuals generation in an industry and to estimate the quantitative responses to variations in those factors

2. to determine the range of options available in a given industry to respond to increasingly stringent constraints placed on the discharge of residuals to the environment, that is, constraints on the use of common property resources as inputs to production processes, and, as a corollary, to estimate how industrial operations will respond to different REQM policies

3. to quantify the proportion of total production costs represented by net residuals management costs, under increasingly stringent constraints on residuals discharges and in relation to different sets of factor input costs (such as fuel and raw materials) and production variables (such as technology of production and product output specifications)

4. to determine the extent to which the physical, technological, and economic interrelationships among the types and states of residuals require that all residuals be considered simultaneously to determine the optimal residuals management strategy for an industrial plant

[4] Clifford S. Russell, *Residuals Management in Industry: A Case Study of Petroleum Refining* (Baltimore, Johns Hopkins University Press for Resources for the Future, 1973).

[5] Clifford S. Russell and William J. Vaughan, *Steel Production: Processes, Products and Residuals* (Baltimore, Johns Hopkins University Press for Resources for the Future, 1976).

[6] Blair T. Bower, George O. G. Löf, and W. M. Hearon, "Residuals Management in the Pulp and Paper Industry," *Natural Resources Journal* vol. 11, no. 4 (1972) pp. 605–23; and George O. G. Löf, W. M. Hearon, and Blair T. Bower, "Residuals Management in Pulp and Paper Manufacture," in Walter S. Kaghan, ed., *Forest Products and the Environment*, American Institute of Chemical Engineers Symposium Series no. 133, vol. 69 (1973).

[7] James W. Sawyer, Jr., *Automotive Scrap Recycling: Processes, Prices, and Prospects* (Washington, D.C., Resources for the Future, 1974).

[8] Jerome K. Delson, Richard J. Frankel, and Blair T. Bower, "Residuals Management in the Coal-Electric Energy Industry," unpublished study.

5. to adapt the detailed industry models for use in analyses of regional REQM. The relevance of this issue is discussed in the description of the Lower Delaware Valley case study in chapter 7 of the present book.

Although the work in the Quality of the Environment program has concentrated on industrial operations, the same approach is relevant for agricultural, silvicultural, mining, commercial, and residential activities.

Conceptual Framework and Analytical Methods

Various conceptual frameworks and analytical methods could be used in studying residuals management in industry (and in other activities). The evolution and essential features of the conceptual framework adopted and analytical approaches used are described below.

A production activitity—manufacturing, mining, logging, agriculture —operates on one or more raw materials via physical, chemical, and biological transformations by use of capital equipment and inputs of human and nonhuman energy to produce one or more desired outputs. However, no production process can be designed for 100 percent conversion of inputs into desired outputs. Thus there are material and energy outflows in addition to the desired outputs of products and energy. The former are termed "nonproduct outputs" of the production activities, for reasons which will be evident in subsequent discussion. They consist of: (1) nonproduct materials *formed* in the production process, (2) raw materials not transformed in the production process such as catalysts, and (3) nonused or nondesired energy outputs from the production process.

It is assumed that the objective at the plant is, at least, loosely, to maximize profits or minimize costs in relation to prices of inputs and outputs and subject to whatever constraints are relevant. Even if there are no constraints on the use of common property resources (such as the atmosphere, biosphere, water bodies), it is economical in many cases to recover and reuse substantial portions of the nonproduct outputs, both material and energy. (Although the discussion immediately below is couched in terms of materials, it is equally relevant to energy flows.)

The extent to which materials recovery is practiced at a particular plant is a function of the relative costs of recovered materials versus new (makeup) materials, whether the latter are purchased on the open market, from another section of the plant, or from another plant of the same company. These relative costs are in turn affected by the technology of the production process, the technology of materials recovery, and the technology of production of the "new" material inputs. Tradeoffs are possible between the design of the production process to reduce the formation of nonproduct materials and the extent of utilization of materials recovery technology. In effect, the combination of the production process plus the materials recovery system is optimized, in the absence of constraints on residuals discharge. With constraints of one type or another imposed on residuals discharge, the total system is optimized—production process, materials and energy recovery, and residuals management. All of the factors cited above are time dependent; hence, the extent of materials and energy recovery at a plant is likely to change over time.

Although essential in terms of describing the ground rules for studies of residuals management in industry, the above provides no operational framework. A first attempt to become operational was based on a formulation adapted from studies of industrial water utilization.[9] Thus, residuals generation in the absence of constraints on residuals discharges is expressed as[10]

$$R_G = f(M, TP, PO, POS) \tag{1}$$

where

R_G = vector of residuals generated per unit of product output or per unit of raw material processed

M = type, and hence characteristics, of raw materials used

TP = technology of production process, including technology of materials and energy recovery and technology of by-product production

PO = product mix

POS = specifications for each of the products in the product mix.[11]

There are other variables that may be of greater or lesser importance, particularly for existing plants, such as climate and the physical layout of the plant. The latter, for example, affects the costs of in-plant modifications, such as water recirculation.

[9] Blair T. Bower, "The Economics of Industrial Water Utilization," in Allen V. Kneese and Stephen C. Smith, eds., *Water Research* (Baltimore, Johns Hopkins University Press for Resources for the Future, 1967) pp. 143–73.

[10] The formulation does not indicate the normally large variability in residuals generation discussed below.

[11] For illustrations of the distinction between *PO* and *POS,* see chapter 3, footnote 1.

Residuals *discharge* into the various environmental media is then a function of the same factors plus the effluent controls imposed on the plant and the technology of residuals modification. Thus,

$$R_D = f(M, TP, PO, POS, EC, TR) \tag{2}$$

where

R_D = vector of residuals discharged per unit of product output or per unit of raw material processed

M, TP, PO, POS are the same as in (1)

EC = controls, standards or charges, or both, imposed on discharge of liquid, gaseous, solid, and energy (heat and noise) residuals

TR = technology of residuals modification.

Of course, the effluent controls may result in shifts in $M, TP, PO,$ and POS.

However, these formulations are not complete enough to provide an adequate conceptual framework to address the objectives stated previously. They are particularly deficient in failing to make explicit the role of: prices of factor inputs (chemicals, water, electrical energy, and heat); various exogenous variables such as tax policies and import quotas; technological changes in other production processes that utilize the outputs of the production process under consideration; and the factors that influence final demand (both PO and POS). Russell has proposed an excellent conceptual model that includes such factors; it is shown in figure 6.[12] Even though it is only a qualitative framework, it serves two important purposes: first, it focuses attention separately on the factors outside the plant which indirectly, and those which directly, affect residuals generation. Second, it emphasizes that within the production process itself other inputs can frequently be substituted which change residuals generation. That is, the model illustrates the fundamental sense in which it is misleading to assume that analyses based on fixed coefficients (such as pounds of BOD_5 generated per unit of output) are conceptually valid.

This conceptual framework was implemented in the Quality of the Environment industry studies by models that involved different combinations of focus and analytical approach, as indicated on page 60.

In the system focus, "system" refers to a set of spatially separate units required to produce a final product output.

[12] Clifford S. Russell, "Models for Investigation of Industrial Response," *Swedish Journal of Economics* vol. 73, no. 1 (1971) pp. 134–156.

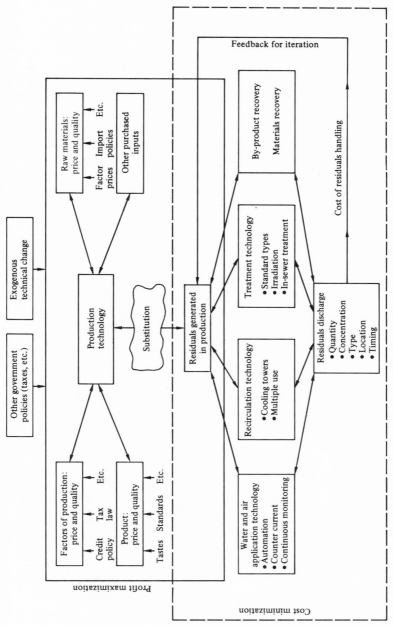

Figure 6. A proposed model of industrial residuals generation and discharge. From: Clifford S. Russell, "Models for In-vestigation of Industrial Response," *Swedish Journal of Economics* vol. 73, no. 1 (1971) p. 137.

	Analytical Approach	
Focus of model	*Response surface sampling*	*Mathematical programming*
Industrial plant	Pulp and paper	Petroleum refining Steel
System	Coal-electric energy	Steel scrap

Thus in the study of petroleum refining a single refinery was analyzed, taking the alternative types of crude oil inputs as given and not associating those inputs with any of the residuals problems involved in providing those inputs, that is, in the activities of exploration, drilling, and transport to the refinery via tanker or pipeline. In contrast, the coal-electric energy industry study considered residuals generation and management throughout the entire system from the coal in the ground to energy produced at the high side of the busbar.

All industry studies except that of the steel industry assumed a grass roots plant, that is, one to be built, not one already in existence. This does not mean that much of what has been learned, including relative costs of residuals management in proportion to total production costs, is not relevant or valid for existing plants. What it means is that the results must be interpreted carefully, because the costs of residuals management for *existing* plants are likely to be particularly site specific, especially with respect to the physical layout of the plant, the existing technology of production, climate, and land costs.

All of the studies required the calculation of materials and energy balances for each of the processes and operations involved in the plant or system. These calculations were based on steady-state conditions. All of the residuals must be estimated in order to ensure consistency in the total materials and energy balances. However, not all residuals must be explicitly modified before discharge, for example, heat, water vapor, and carbon dioxide discharged into the atmosphere. However, in order to size and cost the residuals modification equipment, the water vapor and carbon dioxide must be calculated where there are other materials in the discharge streams slated for modification.

Variability in Residuals Generation and Discharge

In the engineering sense, a given production activity, such as the manufacture of paper, is designed to produce a range of output types, that is, grades of paper. One particular grade is likely to be dominant in terms of proportion of total output, so that the plant is designed to achieve maximum production efficiency—in the physical sense of converting raw material into product output—when producing this grade.

Once a production facility is in operation, substantial variations in residuals generation (R_G) and discharge (R_D) occur over time because of a number of factors. At least three types of variation can be identified according to whether they take place: (1) during normal production operations, (2) during start-up and shut-down operations, or (3) during breakdowns and accidental spills. Variations during normal production operations can be subdivided into: (a) less than daily, and daily variations; (b) less than seasonal production cycle variations, reflecting changing product mix; and (c) seasonal variations.[13] Only variability during normal production operations is discussed here.

The major factors resulting in less-than-daily (within the day) and daily (day-to-day) variations are: (1) variation in quantity or quality, or both, of raw material input, hence, variation in operating rate (units processed or produced per unit of time); (2) variation in operating conditions, that is, temperatures and pressures; (3) variation in conditions of operating equipment; (4) variation in product mix. The following are examples of each:

1. Typically the inflow of raw tomatoes to a cannery varies within a given day in quantity and quality because of harvesting and transport conditions. The quality may vary because of different conditions at the different sources of raw tomatoes for the cannery.
2. Operating conditions in a production process may change in a particular pattern over a production cycle of a production process, where such a cycle may be a few hours or a few weeks. For example, basic conductivity in steelmaking changes throughout the cycle of a basic oxygen furnace. Therefore R_G changes during the cycle and hence, for a given installation of particulate removal equipment, R_D varies throughout the cycle, because removal efficiency is a function of conductivity. A not atypical situation is for a production unit to be "pushed," that is, operated at higher than design capacity. Power

[13] These subdivisions are not perfectly unambiguous. Some component of the less-than-daily variation is a function of the same changing product mix variable that yields the seasonal and production cycle variations.

plants often operate at 105–110 percent of design capacity for short periods.

3. In paper production, the efficiency of the saws and chippers changes with use, so that R_G gradually increases over time, until the saws and chippers are resharpened. Similarly, the effectiveness of the felts on a paper machine gradually decreases over time, thereby increasing R_G.

4. Product mix may change within a single day, as well as from day to day. In a medium or large tomato cannery, the mix of canned tomatoes, tomato paste, tomato juice, and so forth, is likely to vary almost from hour to hour. Similarly, a large paper mill with several paper machines, each producing a certain range of products, will probably show substantial variations in product mix within a single day. Even where a mill is producing only one product, for example, linerboard, substantial variation occurs in R_G stemming from the production of different grades of linerboard, often in a single day. A small paperboard plant, that is, one producing about 100 tons per day from wastepaper, may well show substantially less-than-daily and daily variations over a typical production cycle of two weeks, beginning with lighter weight board and producing increasingly heavy grades to the end of the cycle.

Variations in R_G over the production cycle gradually evolve into seasonal variations. Climate is one factor yielding seasonal variations; change in production (output, mix, or both) is another. For some production operations there is a definite seasonal pattern, for example, in: a large cannery packing various fruits, vegetables, soups, where R_G depends on which raw product is in season; a petroleum refinery producing both gasoline and fuel oil where the proportion of the two varies with the season (winter versus summer); a fish cannery or beet sugar plant operating only during the period of availability of raw product; the skewed production cycle of automobile assembly plants. Many activities other than manufacturing exhibit major variations by season, such as agricultural operations, logging operations, resort operations, and household activities, with resultant time variation in R_G. In general, variation in R_G is larger for gaseous residuals than for liquid residuals.

The variation in R_D may be less than, about the same as, and in some cases even greater than, the variation in R_G, depending on the modification

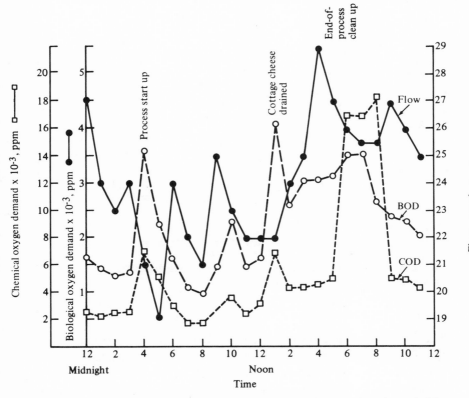

Figure 7. Hourly variation in residuals generation for a dairy plant: BOD$_5$, COD, and wastewater flow

process involved. If the process is sensitive to such factors as input loading and temperature, the resulting variation in R_D from day to day may be significantly greater, even though the median R_D is much less than the median R_G.

Figures 7 and 8 illustrate variability in two types of operations—a dairy plant and a paper mill, respectively. The latter has a biological modification process following sedimentation of the primary liquid residual. Some of the implications of variations in R_G and R_D are discussed in chapter 10. We now consider selected results from the RFF Quality of the Environment industry studies.

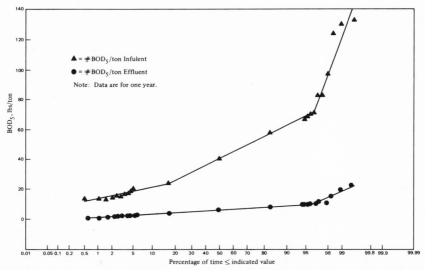

Figure 8. Frequency distribution of daily residuals generation and residuals discharge for an integrated paper mill

The Pulp and Paper Industry

The types and quantities of residuals generated in the production of a specific type of paper depend primarily on the raw materials employed, the pulping process used, the extent of bleaching, and the characteristics desired in the paper product. All of these factors are of course interrelated. The desired brightness of the final product determines the amount of bleaching needed for a given raw material and pulping process. Similarly, the desired strength—or any other product characteristic— limits the combinations of raw materials and pulping processes that can be used.

To give specificity to these general statements, consider an integrated paper mill producing jumbo rolls of tissue paper.[14] (The discussion excludes the converting operation leading to the final user product.) Table 2 shows the residuals generated in producing one ton of tissue paper for different combinations of raw material, pulping process, bleaching, and

[14] An integrated paper mill is one that produces both pulp and paper. The converting operation, that is, transforming the jumbo roll of paper into the packages of tissue paper as purchased by the final user, does not affect the materials and energy balances of the unit processes and operations. Rather, the total amount of throughput is affected, depending on the amount of loss in converting and the disposition of the converting residuals.

TABLE 2. Residuals Generated in Producing One Ton[a] of Tissue Paper (pounds per ton)

	Brightness (General Electric brightness [GEB] 80–82)							
				Production combination (P.C. No.)				
				Pulp mix				
Residual	Ti 1 100% Ca/CEH	Ti 2 100% NH₄/CEH	Ti 3 100% Mg/CEH	Ti 4 100% K/CEHD	Ti 5 50% K/CEHDED; 50% SG/Zn	Ti 6 50% Mg/CEHH; 50% SG/Zn	Ti 7 50% K/CEHD; 50% Mg/CEH	Ti 8 75% K/CEHDED; 25% WPN/FIB
Gaseous								
Chlorine	1.1	1.1	1.1	1.2	0.6	0.6	1.1	0.9
Chlorine dioxide	0	0	0	0.6	0.6	0	0.3	0.9
Sulfur dioxide[b]	125/34.0	114/29.0	48.7/15.0	5.6/20.0	2.8/25.0	24.6/23.0	27.1/17.0	4.2/19.0
Hydrogen sulfide and organic sulfide	0	0	0	25.5	12.8	0	12.7	19.1
Particulates[b]	27.5/1.7	27.1/1.4	27.7/0.8	57.5/1.0	34.6/1.3	19.8/1.2	42.6/0.9	43.2/1.0
Liquid								
Dissolved inorganic solids[e]	127	130	108	263	159	96	185	244
Dissolved organic solids[e]	2970	2900	190	244	193	178	217	261
Suspended organic solids	113	112	109	113	97	100	111	139
Suspended inorganic solids	4.4	4.4	4.1	4.5	3.2	3.0	4.3	3.3
5-day biochemical oxygen demand	820	804	92	147	105	84	120	151
Solid								
Inorganic solids	55.4	46.0	77.9	82.0	50.3	49.7	79.9	65.1
Organic solids	83.0	63.1	63.1	0	0	31.9	31.5	3.2

TABLE 2. (continued)

	Ti 21	Ti 22	(70–72 GEB) Ti 23	Ti 24	Ti 25	Ti 31	(25 GEB) Ti 32	Ti 33	Ti 34
	100% Mg/H	100% K/CEH	50% Mg/H; 50% Sg/Zn	60% K/CEHD; 40% WPM/FIB	35% K/CEHD; 65% WPN/FIB	100% K/O	50% Mg/O; 50% SG/O	35% K/O; 65% WPN/F	100% WPM/F
Gaseous									
Chlorine	0	1.2	0	0.7	0.4	0	0	0	0
Chlorine dioxide	0	0	0	0.3	0.2	0	0	0	0
Sulfur dioxide	48.1/25.0	5.6/20.0	24.0/23.0	3.4/20.0	2.0/19.0	5.1/7.0	23.4/19.0	1.8/17.0	0/17.0
Hydrogen sulfide and organic sulfides	0	25.4	0	15.3	8.9	23.2	0	8.1	0
Particulates	27.4/0.7	57.4/1.0	19.5/1.1	34.5/1.0	20.1/0.9	52.4/0.3	19.1/1.0	18.4/0.6	0/0.8
Liquid									
Dissolved inorganic solids	103	263	74	235	192	22	17	15	21
Dissolved organic solids	144	227	140	311	278	41	108	29	63
Suspended organic solids	108	113	94	145	178	107	93	105	92
Suspended inorganic solids	4.1	4.4	3.0	101	1.6	4.1	2.9	1.4	202
5-day biochemical oxygen demand	79	143	69	163	148	31	46	27	36
Solid									
Inorganic solids	79.0	81.8	48.7	55.6	37.7	73.7	44.1	32.8	13.8
Organic solids	62.2	0	31.1	6.0	8.3	0	30.3	6.9	12.8

Notes: Raw material is softwood logs, except where waste paper is used.

Specifications other than brightness: 14 basis weight (single ply)

low wet strength (25%)

Abbreviations: Ti = tissue paper; Ca = calcium base sulfite pulping; NH_4 = ammonium base sulfite pulping; Mg = magnefite (sulfite) pulping; K = kraft (sulfite) pulping; SG = stone groundwood pulping; WPN = wastepaper, No. 1 News (raw material); WPM = wastepaper, No. 1 Mixed (raw material); CEH, CEHD, etc. = kraft or magnefite bleaching sequences where C = chlorination, E = caustic extraction, H = hypochlorite bleaching, and D = chloride dioxide bleaching; Zh = groundwood bleaching, zinc hydrosulfite; O = no bleaching; F = wastepaper processing-defibering; FIB = wastepaper processing-defibering, drinking, and bleaching.

[a] Output is as air-dry paper, equivalent to 1,880 pounds on a bone-dry basis.

[b] 1% sulfur fuel oil is assumed for the purchased fuel used to generate heating steam and electric energy for plant use. **Right-hand figure in each column of these rows is the quantity generated associated with fuel combustion.**

[c] The division of total dissolved solids into organic and inorganic portions is to a partial extent arbitrary. For example, a dissolved compound comprised of a metal and a wood ingredient or derivative might be considered organic, inorganic, or partially in each category. In the analyses represented in this table, most of the dissolved excess and wasted chemical agents are classified as dissolved inorganic solids; most of the dissolved organic fractions of wood or wastepaper, as dissolved organic solids.

product brightness. Not shown are by-products and normally innocuous nonproduct outputs of water, consisting of carbon dioxide and nitrogen. The residuals generated in the combustion of purchased fuel used in generating process steam and electric energy are included.

The table illustrates clearly the effects of the variables, for example: production process—Ti 3 (magnefite)[15] versus Ti 4 (kraft); product brightness—Ti 31 versus Ti 4; type of raw material—Ti 34 versus Ti 31. For these comparisons, respectively: (1) the magnefite process generates no reduced sulfur compounds and less than half the particulates, but almost 2½ times the SO_2 compared with the sulfate process; (2) reducing product brightness and hence the extent of bleaching—all other specifications remaining the same—from 80 to 25 (the brightness of unbleached kraft) cuts SO_2 in half, dissolved solids by over 85 percent, and BOD_5 by almost 80 percent; (3) using No. 1 mixed wastepaper as the raw material instead of softwood logs results in no reduced sulfur compounds, essentially no particulates, an increase of almost 50 percent in SO_2, some increase in dissolved solids, and many times more suspended inorganic solids. For other types of wastepaper, the quantities would of course be different.

The data in table 2 are based on materials, energy, and heat balances for each of the processes and operations in the different production combinations, including on-site fuel combustion to generate process steam and energy. They represent the sums of all of the individual streams of each residual generated, after taking into consideration the amounts of materials recovery and recirculation of water and heat estimated to be economical at the time (1970), in the absence of constraints on residuals discharges. The balances used for production process Ti 4 are shown in the flow sheets in appendix 4-B.

The individual residuals streams are the basis for the analysis of residuals management at the plant level. This will be illustrated by considering production process Ti 4 for an integrated kraft mill producing 500 tons per day of tissue paper. Producing 500 tons per day of *paper* requires a pulping capacity of about 580 tons per day. Other unit processes and operations are sized as necessary to enable production of the specified output. Table 3 shows the significant sources of particulate generation for the mill. Most of the nonproduct particulates generated in the recovery furnace and lime kiln are recycled to the chemical system. The quantities shown in the table are those in excess of the economically recoverable amounts.

[15] Eighty percent of spent liquor was estimated to be economically justified.

TABLE 3. Main Sources of Particulate Generation in an Integrated Kraft
Paper Mill

Source	*Pounds of particulates* *generated per ton of paper*
Recovery furnace stack	5.0
Lime kiln stack	11.7
Bark boiler stack	27.6
Slaker vent	0.7
Smelt-dissolving tank vent	1.2
Fuel-fired boiler stack	1.0[a]
Total	47.2

[a] Particulates from combustion of oil for generation of steam and electric energy.

Figure 9 shows the net annual cost for modification of the nonproduct particulate materials formed in the flue gases from the recovery furnace, the first source listed in table 3.[16] The points shown above the curve represent alternative measures for the same degree of particulate modification, but with higher costs (all costs in this analysis are in terms of 1970 dollars). Up to the level of modification designated by X, particulate modification would be undertaken in the absence of effluent controls because the value of the recovered chemicals is larger than the cost of recovery. Up to recoveries of about 90 percent, electrostatic precipitators of moderate size are assumed to be used. Greater than 90 percent recovery is accomplished by use of larger precipitators and the addition of cyclonic scrubbers or venturi scrubbers, following precipitators. In this case the materials recovery–residuals modification technology involves end-of-pipe measures throughout the entire range of handling the nonproduct outputs.

Similar relationships were developed for each of the particulate streams. Figure 10 shows the relationships for the bark boiler stack, the slaker vent, and the smelt tank vent. For the bark boiler stack, up to 85 percent removal is assumed to be achieved by a single cyclone separator, to 95 percent in multiclones, and above 95 percent in two-stage multiclone units. Seventy-five percent of the particulates in the slaker vent are removed by

[16] Costs are illustrative. Variation in gas flow rates, particle sizes, equipment costs, amortization schedules, and other factors may result in significant differences in costs from plant to plant for the same raw material-process-product combination. All cost estimates for the pulp and paper study were adjusted to a 1970 basis. Net annual cost is comprised of the annual costs of energy, supplies, operating labor, maintenance labor, plus fixed costs at an annual rate of 12.5 percent of capital investment, net of the value of recovered material and energy. Operation 350 days per year was assumed.

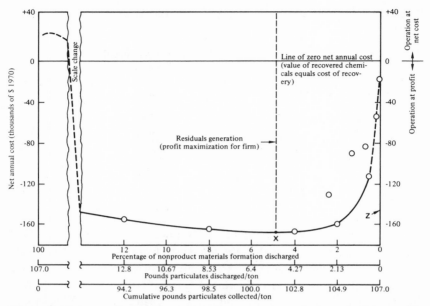

Figure 9. Net annual cost of recovery furnace nonproduct particulate modification: integrated kraft mill producing 500 tons per day of bleached tissue paper. From: George O. G. Löf, W. M. Hearon, and Blair T. Bower, "Residuals Management in Pulp and Paper Manufacture," in Walter S. Kaghan, ed., *Forest Products and the Environment,* American Institute of Chemical Engineers Symposium Series, no. 133, vol. 69 (1973) p. 146 (reprinted by permission of the senior author).

mesh pads and 85 percent by a cyclone scrubber. For the dissolving tank vent, 75 percent removal is achieved by mesh pads, 90 percent by a packed tower, and 97 percent by an orifice or a venturi scrubber. In contrast to the situation represented in figure 9, no level of particulate removal is economically justified for the bark boiler stack and the slaker vent. For the smelt tank vent, about 75 percent recovery is economically justified in the absence of constraints on residuals discharges. Subsequent analysis of liquid residuals modification indicated that the incremental costs associated with having to modify secondary liquid residuals from wet scrubbers were high enough to cause a shift to "dry" gaseous residuals modification measures for all particulate residuals streams.

Table 4 shows the results of combining the analyses of the individual particulate streams to determine the minimum cost to meet specified particulate discharge standards.

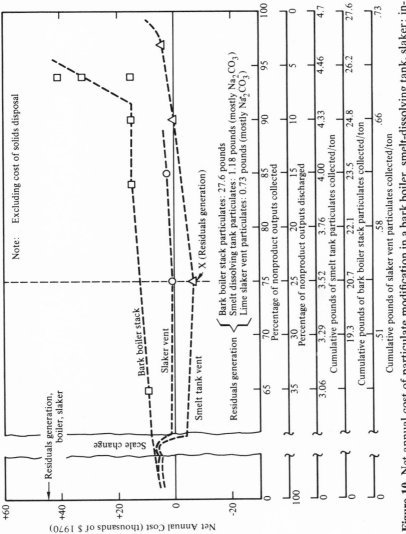

Figure 10. Net annual cost of particulate modification in a bark boiler, smelt-dissolving tank, slaker: integrated kraft mill producing 500 tons per day of bleached tissue paper. From: Löf, Hearon, and Bower, "Residuals Management," p. 147 (reprinted by permission of the senior author).

TABLE 4. Net Annual Cost of Particulate Modification in an Integrated Kraft Mill Producing 500 Tons per Day of Bleached Tissue Paper

Item	No modification		Low modification		Medium modification		High modification	
	Pounds of particulates/ton	Annual net cost (dollars)	Pounds of particulates/ton	Annual net cost (dollars)	Pounds of particulates/ton	Annual net cost (dollars)	Pounds of particulates/ton	Annual net cost (dollars)
Particulate discharge	47.0		30		8		4	
Particulate removal								
Bark boiler stack	0	0	17.2	8,940	25.9	14,300	26.0	19,500
Lime kiln stack	0	0	0	0	10.3	17,000	11.6	32,900
Recovery furnace stack	0	0	0	0	2.5	5,000	3.9	32,400
Lime slaker vent	0	0	0	0	0.5	750	0.6	2,390
Smelt tank vent	0	0	0	0	0	0	1.0	10,560
Fuel boiler stack	0	0	0	0	0	0	0	0
Total removal	0	0	17.2	8,940	39.2	37,050	43.2	97,750
Cost of solids disposal		0		3,150		4,540		4,560
Cost of liquid residual disposal		0		0		0		0
Total net annual cost of particulate control		0		12,090		41,590		102,310

Notes: Costs are based on estimates of operating labor, maintenance labor and supplies, energy and material requirements, 12.5 percent annual fixed cost on estimated capital investment, and credited with chemical recoveries at typical prices. Adjustments to 1970 costs were made. Operation 350 days per year was assumed. The gas streams to be modified and the extent of particulate removal from a stream were selected so as to obtain the lowest cost of removing the required quantity of particulates regardless of particulate composition. Cost of solids disposal is only for the solids resulting from removal of particulates not recycled in the mill, and excludes disposal costs for other solid residuals. It is based on a nominal $2 per ton hauling and landfill cost. None of the particulate removal processes results in generation of a liquid residual. Costs are in 1970 dollars.

Source: Löf, Hearon, and Bower, "Residuals Management," pp. 141–49 (reprinted by permission of the senior author).

TABLE 5.   Specification of Increasingly Stringent Discharge Standards in an Integrated Kraft Mill Producing 500 Tons per Day of Bleached Tissue Paper (lbs/ton)

	Level of discharge standards				
Residual	0	I	II	III	IV
SO_2	No control[a]	50	35	20	10
Particulates	No control	30	8	4	2
Reduced sulfur compounds	No control	10	2	0.5	0.2
Suspended solids	No control	50	20	10	5
BOD_5	No control	60	35	20	10

Note: Discharge standards apply to total mill operation, that is, from all sources.

[a] No restrictions on discharges; reflects basic production costs.

Similar analyses were made for the various nonproduct liquid streams. One obvious point should be noted, namely, that minimum-cost liquid residuals modification measures often involve combining various streams of the same residual for simultaneous (joint) modification. Rarely is it economical to combine several streams of the same gaseous residual generated at different locations in the plant.

To demonstrate the effect of increasingly stringent effluent controls, four sets of discharge standards were posited involving the major residuals. These sets are shown in table 5.

Net residuals management costs per ton of output to meet the sets of standards in table 5 are shown in table 6 for the same type of integrated kraft mill that uses the production process P.C. Ti 4. The much higher costs for liquid residuals modification than for gaseous residuals modification for this production combination reflect the substantial value of materials recovered from gaseous residuals streams and the fewer such possibilities existing for the liquid residuals streams. Also shown in table 6 for comparison, are the net residuals management costs per ton for a paper mill of the same output capacity but producing *unbleached* tissue paper (P.C. Ti 31). The impact of changing just one product specification is very significant, residuals management costs for unbleached tissue paper being about 20 to 25 percent of those for bleached paper. This is illustrated graphically in figure 11, which shows total rather than marginal costs. Also illustrated in that figure are the effects of economies of scale, which exist for many types of residuals modification options.

All costs developed in the pulp and paper industry study include the costs of energy used in handling, modifying, and disposing of residuals. The costs also take into account any changes in energy, material, and

TABLE 6. Net Residuals Management Costs per Ton of Output in an Integrated Kraft Mill Producing 500 Tons per Day of Tissue Paper (1970 dollars per ton)

Item	Level of discharge standards			
	I	II	III	IV
P.C. Ti 4: bleached tissue paper				
Gaseous residuals modification	0.16	0.69	1.77	3.67
Liquid residuals modification	3.08	4.10	6.76	10.47
Solid residuals disposal	0.38	0.38	0.38	0.38
Total	3.62	5.17	8.91	14.52
P.C. Ti 31: unbleached tissue paper				
Gaseous residuals modification	0.12	0.51	0.92	1.97
Liquid residuals modification	0.21	0.54	0.87	1.49
Solid residuals disposal	0.33	0.33	0.33	0.33
Total	0.66	1.38	2.12	3.79

Notes: Costs are in 1970 dollars and are based on estimates of operating labor, maintenance labor and supplies, energy and material requirements, 12.5% annual charge on estimated capital investment, and are credited with chemical recoveries at typical market prices. Operation at 350 days per year was assumed; the costs of any secondary solid residuals generated in liquid and gaseous residuals modification, e.g., sludge, are included in the liquid and gaseous residuals modification costs.

capital costs of production resulting from the adoption of a particular residuals modification option, such as the substitution of an indirect for a direct contact evaporator.

The major paper products (linerboard, corrugated paper, medium-strength papers, boxboard, consumer products, newsprint, and printing papers); the major pulping processes (kraft, sulfite, neutral sulfite semi-chemical, stone and refiner groundwood, and hydrapulping of waste-paper); and alternative raw materials (round wood, wood products residues, and wastepaper) were studied. Residuals management costs were estimated for mills using various combinations of raw material, pulping and bleaching processes, and product output, as illustrated above for the softwood-kraft-tissue paper production combination. Many paper mills actually have a mix of combinations of raw material and pulping and bleaching to produce multiple products. Often different types of pulp are combined to produce a given product, depending on raw material availability, installed pulping capacity, and characteristics of the desired product. Where multiple processes and products are involved in a single plant, residuals management becomes more complicated because the number of options for handling residuals increases substantially. Residuals management costs are usually not simply the sum of those for the individual production combinations.

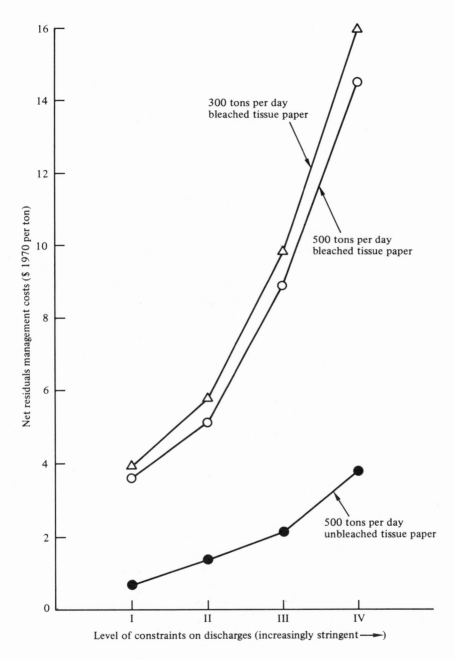

300 tons per day
bleached tissue paper

500 tons per day
bleached tissue paper

500 tons per day
unbleached tissue paper

Level of constraints on discharges (increasingly stringent ⟶)

Figure 11. Net residuals management costs for an integrated kraft mill.

Petroleum Refining

The study of residuals management in petroleum refining used a linear programming model that, once constructed, permitted rapid analysis of the effects on residuals discharges, on costs of various factors affecting residuals generation, and on different restrictions on residuals discharges. The study focused on a 150,000-barrel-per-day (crude charge) "grass roots" refinery with the size of process units and magnitudes of product outputs shown in table 7. This is a small new refinery. Only a brief review of a few of the results is reported here.[17]

Table 8 summarizes the changes in residuals discharged in response to changes in a number of direct and indirect influences. The table shows: (1) percentage changes in residuals discharged, total heat applied, and by-product sulfur produced; and (2) changes in cost or gain per barrel of crude processed, and percentage changes in cost resulting from changes in type of refinery,[18] product mix, sulfur content of crude, prices of water withdrawals and fresh heat, and the imposition of a $0.05 per lb effluent charge on residuals discharged—phenols, BOD_5, amonia (NH_3), sulfur dioxide (SO_2), and particulates. The results demonstrate clearly that indirect influences are important and should not be ignored in analyzing residuals management in industry. For example, a change in gasoline quality specifications produces changes in residuals discharges of the same order of magnitude as the imposition of a significant effluent charge.

The table supports the following observations:

1. Certain indirect or exogenous influences have very large effects on all or most of the residuals discharges. This is true of water costs, product quality specifications, and product mix requirements. The effects of crude sulfur content are only slightly smaller and less widespread.

2. Changing the cost of water withdrawals (and pretreatment) affects the liquid residuals discharges dramatically, but has no impact on discharges to the atmosphere. In contrast, the effects of changes in the price of fresh heat are almost all confined to the SO_2 and particulate residuals.

3. Although direct comparisons of the costs implied by the various indirect influences are made difficult by the very nature of the changes specified, a few are possible. Clearly, a tripling of the cost of fresh heat

---

[17] A detailed description is found in Clifford S. Russell, *Residuals Management in Industry: A Case Study of Petroleum Refining* (Baltimore, Johns Hopkins University Press for Resources for the Future, 1973). Costs cited are in 1970 dollars.

[18] The "advanced" refinery incorporates an advanced hydrogen-intensive cracking process which enables a higher gasoline yield from the given crude mix and more flexibility in the composition of the output.

TABLE 7. Process Units, Product Outputs, Costs, and Prices for a 150,000-Barrel-Per-Day Petroleum Refinery[a]

Process units ($ per barrel/crude charged)		Product outputs, bench mark case (quantity per day)	
Atmospheric distillation	1.00	Products sold	
Desalter	1.00	Refinery gas (lbs)	2.944×10^6
Coking	0.133	Kerosene/diesel oil	15,760
Hydrotreating	0.139	Distillate fuel oil (barrels)	
Reforming	0.139	Low sulfur	8,880
Catalytic cracking	0.466	Medium sulfur	8,230
Alkylation	0.076	High sulfur	290
Sweetening	0.393	Total	17,400
		Polymer (barrels)	660
		Premium gasoline[b] (barrels)	35,100
		Regular gasoline[c] (barrels)	51,150
		Residual fuel oil (barrels)	3,000
		Straight-run gasoline sold as petrochemical feed (barrels)	16,360
		Recovered sulfur (long tons)	40
		Products used internally (lbs)	
		Hydrogen (burned)	100,250
		Sweet coke (burned)	1,180,000
		Sour coke (burned)	260,000
		Coke burned in catalyst regeneration	1,540,000

Costs and prices
(1970 dollars)

Cost of water withdrawals per 1,000 gal.
Cooling	0.015
Desalter	0.070
Process steam	0.15

Cost of purchased fresh heat per 10^6 Btu.
0.5% sulfur	0.661
1.0% sulfur	0.593
2.0% sulfur	0.477

Price of recovered sulfur: 20/long ton

[a] Crude charged: 111,000 barrels East Texas, 0.4% sulfur, at $3.75/bbl
  39,000 barrels Arabian Mix, 1.44% sulfur, at $3.02/bbl
  Total: 150,000 barrels per day.
[b] 100 Octane; average tetraethyl lead content, 2.50 cc/gal.
[c] 94 Octane; average tetraethyl lead content, 2.46 cc/gal.

would be a more serious matter to the refiner than a tripling in water withdrawal costs. Large changes in product mix, including greatly increased production of desulfurized distillate and residual fuel oil, also imply large costs under the assumption of constant product prices. Indeed, it is rather striking to find that all the costs per barrel are of roughly the same order of magnitude.

4. One of the largest effects of requiring a product mix heavy in hydro-desulfurized streams, is the increased recovery of by-product sulfur. Sulfide and ammonia discharges also increase significantly.

5. Effluent charges on different residuals produce widely varying effects on discharges and costs. To facilitate comparability, table 8 is based on a constant $0.05 per pound charge for each of five residuals:

A. When applied to oil and sulfide discharges, this particular charge level has no effect at all; the refiner simply chooses to pay the resulting small fee. The same is true for a $0.05 per billion Btu charge for waste heat discharge. Accordingly, substantial reduction in heat discharges via a charges system would require considerably higher charge levels.

B. For phenols, BOD_5, and ammonia, however, the $0.05 per pound charge is sufficiently high to evoke significant reductions in discharges for most of the liquid residuals. (Increases in particulates discharged result from the incineration of sludge produced in modifying the liquid residuals.)

C. There is no consistent pattern to the relative sensitivity of the basic and the advanced refineries. The basic refinery is more sensitive to the charge on phenols, while the advanced reacts more sharply to the charges on BOD_5 and ammonia.

D. Both refineries respond to a $0.05 per pound charge on particulate discharges with significant discharge reductions; achieving these reductions has no side effects on any other refinery residual (except solids, which are not included here). Total payments for the basic refinery in this situation are somewhat higher than for the liquid residuals.

E. Sulfur dioxide is the refiner's most expensive problem. Faced with a $0.05 per pound charge on its discharge, he opts for very small discharge reductions and pays very large discharge fees, for both basic and advanced refineries.

Further insights into the pattern of response to effluent charges are evident in figure 12. This figure shows the percentage reductions in the discharges of various liquid residuals resulting from increasing levels of charges on BOD_5 discharges, for the basic refinery, benchmark product mix. Note that reductions in discharges of other liquid residuals occur, even though a charge is being levied only on BOD_5.

| | | Percentage changes in residuals discharges | | | | | | | Percentage change in total heat applied |
	Heat to water	Sulfide	Phenol	Oil	BOD	NH₃	SO₂	Particulates	
For basic refinery									
Tripling cost of water withdrawals	−100	−99.2	−84.0	−77.0	−46.1	−31.2	neg.	0	0
Doubling cost of fresh heat ($0.20 to $0.40/10⁶ Btu)	−1.8	0	0	−1.8	0	0	−3.2	+116.0	−4.0
Tripling cost of fresh heat ($0.20 to $0.60/10⁶ Btu)	−2.8	0	0	−2.6	0	0	−4.2	+116.0	−5.6
Increasing average sulfur content of crude by 50% (0.4 to 0.608%)	+1.2	+54.4	+2.6	+1.2	+2.4	+54.8	+17.2	+1.4	+1.2
For advanced refinery									
To modified bench mark product mix with no-lead, low-octane gasoline[b]	+8.0	+33.0	+3.0	+6.1	+1.6	+55.0	+5.4	+0.2	+11.8
To modified bench mark product mix with no-lead, high-octane gasoline[b]	+11.3	+66.7	−8.8	+10.2	−7.9	+90.0	+12.2	−2.6	+22.6
To low gasoline (no-lead, low-octane); high fuel oil; all fuel oils + kerosene desulfurized[d]	−16.4	+167.0	−73.5	−18.4	+1.6	+185.0	−2.7	−68.1	+25.6
Effluent emission charges[e]									
Phenols, basic	−0.4	−11.5	−22.4	−0.3	−14.7	−4.4	0	+0.2	0
Phenols, advanced[d]	0	0	−5.4	0	−1.0	0	0	neg.	0
BOD, basic	0	−23.8	−45.9	−0.4	−42.5	−22.7	0	−0.7	0
BOD, advanced[d]	0	−64.2	−35.5	−0.8	−73.5	−56.9	0	+4.0	0
BOD, no recirculation, basic	0	−13.7	−22.8	−0.1	−20.3	−6.6	0	+0.4	0
NH₃, basic	0	−31.2	−8.8	−0.3	−9.0	−31.1	0	+0.2	0
NH₃, advanced[d]	0	−88.7	−4.7	−0.3	−66.3	−81.5	0	+3.5	0
SO₂, basic	−1.0	0	0	−1.0	0	0	−1.0	−0.1	−1.6
SO₂, advanced[d]	−2.2	0	0	−2.1	0	0	−1.8	−0.5	−2.3
Particulates, basic	0	0	0	0	0	0	0	−82.5	0
Particulates, advanced[d]	0	0	0	0	0	0	0	−69.5	0

TABLE 8. (continued)

	Percentage change in by-product sulfur	Cost or gain ($/bbl crude)			Percentage change in cost
		Cost of discharge reduction	Discharge fee paid	Total cost increase	
For basic refinery					
Tripling cost of water withdrawals	0	—	—	0.031	0.70
Doubling cost of fresh heat ($0.20 to $0.40/$10^6$ Btu)	0	—	—	0.054	1.23
Tripling cost of fresh heat ($0.20 to $0.60/$10^6$ Btu)	0	—	—	0.098	2.24
Increasing average sulfur content of crude by 50% (0.4 to 0.608%)	+46.3	—	—	a	a
For advanced refinery					
To modified bench mark product mix with no-lead, low-octane gasoline[b]	−2.8	—	—	0.076[c]	1.72[c]
To modified bench mark product mix with no-lead, high-octane gasoline[b]	+39.4	—	—	0.206	4.65
To low gasoline (no-lead, low-octane); high fuel oil; all fuel oils + kerosene desulfurized[d]	+88.0	—	—	0.071	1.60
Effluent emission charges[e]					
Phenols	0	0.0002	0.0013	0.0015	0.03
Phenols, advanced[d]	0	neg.	0.0004	0.0004	0.01
BOD, basic	0	0.0007	0.0017	0.0024	0.05
BOD, advanced[d]	0	0.0010	0.0009	0.0019	0.04
BOD, no recirculation, basic	0	0.0003	0.0024	0.0027	0.06
NH$_3$, basic	0	0.0002	0.0007	0.0009	0.02
NH$_3$, advanced[d]	0	0.0010	0.0005	0.0015	0.03
SO$_2$, basic	0	0.0003	0.0707	0.0710	1.68
SO$_2$, advanced[d]	0	0.0005	0.0684	0.0689	1.57
Particulates, basic	0	0.0023	0.0037	0.0060	0.14
Particulates, advanced[d]	0	0.0006	0.0020	0.0026	0.06

Note: Dashes = not applicable. All dollars are 1972 dollars.

Source: Clifford S. Russell, *Residuals Management in Industry: A Case Study of Petroleum Refining* (Baltimore, Johns Hopkins University Press for Resources for the Future, 1973) p. 172.

[a] Cost here depends on the assumed price of high-sulfur crude. At $3.02/bbl, it is $0.109/bbl.

[b] Product mix requirements: premium gasoline 35,100 bbl/day; regular gasoline 51,150 bbl/day; kerosene 16,500 bbl/day; distillate fuel oil 15,000 bbl/day; residual fuel oil 3,000 bbl/day.

[c] Gain instead of cost.

[d] Advanced refinery with product mix requirements; premium gasoline (92 octane) 21,060 bbl/day; regular gasoline (90 octane) 30,690 bbl/day; kerosene (desulfurized) 16,500 bbl/day; distillate fuel oil (low sulfur) 40,500 bbl/day; residual fuel oil (desulfurized) 16,500 bbl/day.

[e] In each case, $0.05/lb. Such charges produced no effects in the cases of oil and sulfide, nor did a $0.05/10^6 Btu charge have any effect on heat discharges.

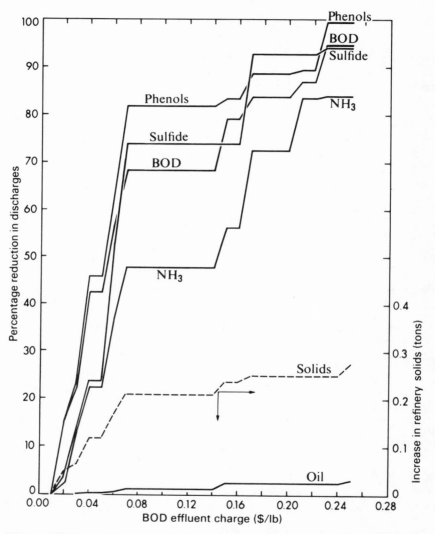

Figure 12. Response to BOD_5 effluent charges: basic refinery, bench mark product mix. From: Clifford S. Russell, *Residuals Management in Industry*, p. 139.

Discharges of BOD_5, sulfide, phenols, and ammonia decrease rapidly over the range of charges from one cent to 7 cents per pound. At the latter level of effluent charge on BOD_5, almost 70 percent of the BOD_5 generation (the load after the oil-water separators and sour-water scrubbers) has been reduced. Note that, simultaneously, the following reductions have occurred in other liquid residuals: phenols—about 80 percent, ammonia—about 45 percent, sulfide—about 75 percent. Reduction of oil is very slight, of heat not at all. No further reduction occurs until the effluent charge is between 14 and 15 cents per pound, when BOD_5 reduction reaches about 80 percent. A charge of about 17 cents per pound induces reduction of BOD_5 of about 90 percent; of 23 cents per pound, about 95 percent.

Reduction in discharge of BOD_5 results in the generation of secondary residuals, particularly "solids" formed in the standard activated sludge process for modifying BOD_5. The assumed rate is 0.75 pounds of dry sludge solids per pound of BOD_5 removed. At an effluent charge of 3 cents per pound of BOD_5, the dry weight of sludge generated is 1,200 pounds per day, in the form of a dilute sludge (5 percent solids). Thus the total weight of raw sludge generated represents about 12 tons per day. At a charge of 22 cents per pound of BOD_5, the weight of raw sludge generated is about 48 tons per day at 5 percent solids. Disposal of this sludge requires thickening and incineration. This in turn results in tertiary residuals of particulates and incinerator bottom ash. The increase in the latter is also shown in figure 12.

This brief discussion of the study of residuals management in petroleum refining illustrates the method of analysis and the useful results that can be obtained from such analysis. Because the linear programming formulation of the industrial residuals management problem is such a powerful and flexible tool of analysis, we present its basic mathematical structure in appendix 4-A at the end of this chapter.

Steel Production

Five major groups of activities are involved in the production of steel: coking, sintering, ironmaking, steelmaking, and final rolling and finishing. For steelmaking itself, there are three basic types of furnaces: open hearth (OH), basic oxygen (BOF), and electric arc (EA). A linear programming model was also used to analyze residuals management in this industry. It included all three furnace types and linked all five groups of activities to produce various types of steel from alternative raw materials with varying costs of the different factor inputs. The study assumed that the basic production facilities, and some residuals modification facilities, were in existence.

The three principal furnace types have distinctive heat sources. In the basic oxygen type, the heat for melting any cold metal charged and for carrying forward the refining reactions is contained in the molten iron charged. For the open hearth, combustion of an outside fuel (oil, coke oven gas, natural gas, and the like) is the heat source. In the electric arc, as the name implies, electrical energy, transformed into heat by an arc, is used. This heat source distinction implies in turn, technological upper limits on the percentage of cold metal (scrap) in the furnace charge and has, under historically prevailing relative costs for ore, scrap, electricity, coal, and other fuels, produced a range of normal charging practices. For the basic oxygen, the upper limit on cold metal input is 30 percent of total metallic input in the absence of such refinements as natural gas lancing for scrap premelting. For the open hearth and electric arc, since the heat source is external, 100 percent scrap may be charged, but, historically, integrated mills (those with blast furnace capacity) have seldom gone below a 50-percent hot metal charge in the open hearth. In contrast, the electric arc is normally operated on a 100-percent cold metal charge, although hot metal may be used.

These differences have important implications for residuals generation and management. The electric arc process is free of the problems associated with by-product recovery, modification, and disposal of residuals generated in coke oven operations, specifically BOD_5, oil, phenols, cyanides, and sulfur. On the other hand, the cold metallic charge to the electric arc can result in a very high level of particulate generation per ton of molten steel when the charge contains a significant portion of No. 2 steel scrap bundles—the proportion possible depending on the type of steel being produced. In addition, the electric arc furnace requires much more electric energy per ton of steel produced, with the consequent generation of larger quantities of gaseous residuals in the associated energy generation than for the open hearth and basic oxygen furnaces.

Each of the furnace types is physically capable of producing, at different costs, three generic types of steel: drawing quality carbon steel (DQ), commercial quality carbon steel (CQ), and alloy steel. The types are defined principally on the basis of the contents of alloy elements—copper, chromium, nickel, molybdenum, and tin. Drawing quality steel has a total alloy content of 0.13 percent; commercial quality, 0.21 percent. The target level of these elements for alloy steel is 1.16 percent, subject to a specified distribution among them. The steel mix investigated in the study exemplifies U.S. production in 1972: 91.5 percent carbon steel and 8.5 percent alloy steel, with 25 percent of the carbon steel assumed to be drawing quality. The mix of drawing quality steel products and the mix of strip thicknesses for each type of steel are arbitrary; it was assumed that thinner gauges should predominate in drawing quality cold-rolled, commercial quality, and alloy hot-rolled, and the reverse for commercial quality and alloy cold-rolled.

Table 9 shows the basic assumptions about the residuals modification equipment in place and about factor input prices. As noted, all costs are in 1968 dollars and factor prices are at 1968 levels. Table 10 shows the pounds of residuals generated per ton of semifinished steel shapes—blooms, billets, and slabs—for a daily output of 2,340 tons, for the three types of furnaces and the specified product mix. The quantities would vary slightly with different mixes of shapes. (Note that any loss in fabrication—that is, converting, if done at the steel mill—is not included.) In addition to the assumptions indicated in the table, it is assumed that only 66 percent of the ammonia produced per ton of coal charged to the coke ovens is contained in the coke oven gas, the remainder being contained in a raw ammonia liquor. Ammonia and phenol recovery from coke plant liquid residuals streams is possible, but would be undertaken only if the market prices of the ammonium sulfate and sodium phenolate by-products are sufficient to cover the costs associated with their production. The prices assumed for the analysis reflected in table 10 result in no recovery being undertaken. Finally, some of the slag generated in the open hearth and basic oxygen steel furnaces could be recycled, depending on the relative costs of processing and disposal and on steel content. No recycling is assumed economically justified in this analysis.

The most pronounced effect of furnace type on residuals generation pertains to particulates. The basic oxygen process results in more than twice as many particulates per ton as the open hearth and electric arc furnaces. (It also generates more of all other residuals per ton except for suspended solids and heat.) However, the scrap prices assumed resulted in a 50-percent hot metal/50-percent scrap charge to the open hearth furnace. Some higher level of absolute scrap prices, with the same ore

TABLE 9.   Base Case Assumptions for Steel Model

Required residuals modification equipment

Process unit	Equipment type	Collection efficiency (percent)
For particulate removal:		
Sinter plant windbox end	Dry cyclone	90
Blast furnace	Dry cyclone plus wet scrubber plus electrostatic precipitator	99
Open hearth furnace	Electrostatic precipitator	97
Basic oxygen furnace	Electrostatic precipitator	94
Electric arc furnace	Fabric filter	97
For suspended solids/oil removal:		
Hot mill:		
Suspended solids		90
Oil		20
Cold mill:		
Suspended solids		30
Oil		0
Blast furnace:		
Gaswasher suspended solids		95.5

Costs and Prices[a]	*$/1,000 gallons*		
Cost of water withdrawals			
Process, once-through cooling	$0.015		
Boiler	0.150		
Hot-rolling mill makeup	0.060		
Cold-rolling mill makeup	0.075		

Scrap prices, 1968	Fe (percent)	*Residual alloy elements* (percent)	$/net ton
#1 Heavy melting	97	0.230	25.97
#1 Factory bundles	99	0.175	29.49
Shredded	97	0.462	25.97
#2 Bundles	85	0.770	19.87
Ore prices ($/net ton, 1968)	55% Fe	65% Fe	66.2% Fe
Pellets	—	—	18.76
Natural (run of mine)	13.97	15.80	—
Coarse	14.73	16.70	—
Fine	13.54	15.29	—

Note: Dashes = not applicable.

[a] Costs are in 1968 dollars; factor prices at 1968 levels.

prices, would induce a 70-percent hot metal/30-percent scrap charge in the open hearth, thereby resulting in substantially higher generation of particulates, and of all other residuals as well, stemming from the corresponding increase in ancillary operations, that is, coke ovens and blast furnace. Particulate generation is also sensitive to type of steel. For example, moving from drawing quality steel to alloy steel, more No. 2

bundles of steel scrap can be used per ton of output because of increasingly higher total alloy content permitted. Number 2 bundles have higher dirt and organic matter, as well as alloy content, compared to other scraps; hence the higher particulate generation per ton as their use increases, the increase being greatest for the electric arc, less for the open hearth, and smallest for the basic oxygen furnace.

It should be emphasized that in these analyses, particulates and SO_2 generated in producing the electric energy required for steel production in each type of mill are included. The energy intensiveness of the electric arc is responsible for the large quantity of SO_2 generated in the electric arc operation, a factor often not reflected in surveys of steel mill residuals generation and discharge.

Of the many other results of the steel industry study that could be cited, only two are selected for discussion here: the effects of continuous casting and the effects of variations in raw material prices. The former illustrate the important role of technological changes in residuals management. The latter illustrate the important effects of changes in relative prices of factor inputs and have implications for dealing with abandoned automobile hulks without considering the technological and economic structure of the only potential (industrial) user of such residuals.

Continuous casting involves the transformation of molten steel directly into semifinished shapes, thus eliminating the traditional ingot forming, ingot reheating, and shape milling. With all three furnace types, installation of continuous casting establishes new relationships within each set of principal subprocesses, especially at the steel furnace. Several factors are involved. First, continuous casting increases yield, that is, reduces total steel to be poured to produce a ton of output. This reduces input requirements at the finishing end. Second, home scrap generation is only about 10 percent of molten steel output with continuous casting versus about 20 percent with conventional casting. Hence, assuming the same scrap-to-hot metal (molten iron) ratio, purchased scrap as a proportion of output increases, and molten iron decreases. Less molten iron means that less blast furnace activity is required, and hence less coking is necessary, with a corresponding reduction of residuals generation associated with coking.

The effects of substitution of continuous for conventional casting are shown in table 11, with all other variables the same as for the bench mark case shown in table 10. All residuals discharges are reduced, except for particulates at the open hearth furnace, the reductions ranging between about 10 and 15 percent for most residuals, except for suspended solids,

TABLE 10. Mill Characteristics and Residuals Discharges by Furnace Type and Coke Quench Practice: Bench Mark Case with Conventional Casting and Full Finishing

	Furnace type				
	Open hearth		Basic oxygen		Electric arc
	Raw ammonia used for quench				
Mill characteristics and residuals discharges	No	Yes	No	Yes	No
Average steel cost ($/ton shape)	70.34	70.34	72.71	72.71	67.77
Marginal iron cost ($/ton hot metal)	40.61	40.61	42.21	42.21	—
Percentage blast furnace capacity used	78.8	78.8	94.8	94.8	—
Scrap purchases (tons/day)					
Pig	—	—	—	—	339.6
#1 Factory	149.9	149.9	0	0	798.6
#1 Heavy melting	436.3	436.3	34.6	34.6	821.4
Shredded	0.0	0.0	0.0	0.0	0.0
#2 Bundle	165.6	165.6	184.5	184.5	200.4
Ore purchased (tons/day)[a]					
High iron content[b]	2,007.2	2,007.2	2,796.9	2,796.9	0.0
Low iron content[c]	80.4	80.4	144.9	144.9	0.0
Coking coal purchased (tons/day)					
1.2% sulfur	0.0	0.0	128.8	128.8	0.0
0.6% sulfur	1,639.4	1,639.4	2,241.5	2,241.5	10.8
Fuel purchased (10^6 Btu/day)					
0.5% sulfur fuel oil	450.6	450.6	0.0	0.0	7,293.4
2.0% sulfur fuel oil	7,186.2	7,186.2	535.4	535.4	15,747.0

	215.2	215.2	253.9	253.9	699.5
Electricity produced (kWh/ton shape)	215.2	215.2	253.9	253.9	699.5
Residuals discharged					
Waterborne					
BOD$_5$ (lbs/ton shape)	5.08	3.32	5.86	3.33	3.32
Oil (lbs/ton shape)	3.17	2.96	3.27	2.96	2.96
Phenols (lbs/ton shape)	0.52	neg.	0.75	neg.	neg.[d]
Cyanide (lbs/ton shape)	0.01	0.0	0.01	0.0	[d]
Ammonia (lbs/ton shape)	1.21	0.0	1.75	0.0	0.01
Suspended solids (lbs/ton shape)	6.23	6.15	5.17	5.05	4.58
Sulfide (lbs/ton shape)	0.19	0.0	0.28	0.0	neg.[d]
Heat (10^6 Btu/ton shape)	5.22	5.22	5.46	5.46	6.31
Airborne					
SO$_2$ (lbs/ton shape)	14.09	14.09	12.31	12.31	16.55
Particulates (lbs/ton shape)	5.91	5.91	12.06	12.06	1.11
Phenols (lbs/ton shape)	0.01	0.53	0.01	0.76	neg.[d]
Land					
Solids (ton/ton shape)	0.14	0.14	0.21	0.21	0.05

Notes: Dashes = not applicable. Neg. = negligible.
[a] Not differentiated by sulfur content or size. Sulfur content assumed to be 0.00015 percent.
[b] 65% iron.
[c] 55% iron.
[d] A very small amount of coking is done for the electric arc in this model. The amounts of residuals involved are correspondingly small and show up only in the third decimal place. Because of this, discharges when quenching is allowed are the same as when it is not.

TABLE 11. Residuals Discharge Variation with the Substitution of Continuous for Conventional Casting[a]

Mill characteristics and residuals discharges	Continuous casting: open hearth	% change from bench mark	Absolute change from bench mark	Continuous casting: basic oxygen	% change from bench mark	Absolute change from bench mark	Continuous casting: electric arc	% change from bench mark	Absolute change from bench mark
Average steel cost ($/ton shape)	66.24	−5.8	−4.10	68.35	−6.0	−4.36	64.28	−5.1	−3.49
% blast furnace capacity used	58.1	−26.3	−20.7	83.2	−18.2	−11.6	—	—	—
Daily iron production (daily tons)	1,302.9	−10.3	−150.2	1,860.7	−12.2	−259.5	—	—	—
Scrap purchases (daily tons)									
Pig	—	—	—	—	—	—	336.3	−1.0	−3.3
#1 Factory	82.8	−44.8	−67.1	45.6	—	45.6	790.2	−1.1	−8.4
#1 Heavy melting	647.0	48.3	210.7	205.8	494.8	171.2	812.5	−1.1	−8.9
#2 Bundle	122.9	−25.8	−47.7	182.5	−1.1	−2.0	198.8	−0.8	−1.6
Ore purchased (daily tons)									
High iron content	1,832.6	−8.7	−174.6	2,458.3	−12.1	−338.6	0.0	0.0	0.0
Low iron content	67.0	−16.7	−13.4	129.9	−10.4	−15.0	0.0	0.0	0.0
Coking coal purchased (daily tons)									
1.2% Sulfur	79.7	—	79.7	113.0	−12.3	−15.8	0.0	0.0	0.0
0.6% Sulfur	1,377.5	−16.0	−261.9	1,967.1	−12.2	−274.4	9.5	0.0	−1.3
Fuel purchased (10^6 daily Btu)									
0.5% Sulfur fuel oil	0.0	−100.0	−450.6	0.0	0.0	0.0	5,027.9	−31.1	−2,265.5
2.0% Sulfur fuel oil	5,065.1	−29.5	−2,121.1	0.0	−100.0	−535.4	13,613.7	−13.5	−2,133.3

	181.1	−15.8	−34.1	214.8	−15.4	−39.1	604.7	−13.6	−94.8
Electricity produced (kWh/ton shape)									
Residuals discharged[b]									
Waterborne									
BOD₅ (lbs/ton shape)	4.27	−15.9	−0.81	4.94	−15.7	−0.92	2.71	−18.4	−0.61
Oil (lbs/ton shape)	2.84	−10.4	−0.33	2.92	−10.7	−0.35	2.65	−10.5	−0.31
Phenols (lbs/ton shape)	0.46	−11.5	−0.06	0.66	−12.0	−0.09	neg.	neg.	neg.
Cyanide (lbs/ton shape)	0.01	neg.	neg.	0.01	neg.	neg.	neg.	neg.	neg.
Ammonia (lbs/ton shape)	1.07	−11.6	−0.14	1.53	−12.6	−0.22	3.59	−21.8	−0.99
Suspended solids (lbs/ton shape)	3.96	−36.4	−2.27	4.11	−20.5	−1.06	neg.	neg.	neg.
Sulfide (lbs/ton shape)	0.17	−10.5	−0.02	0.24	−14.3	−0.04	neg.	neg.	neg.
Heat (10⁶ Btu/ton shape)	4.59	−12.1	−0.63	5.16	−5.5	−0.30	5.56	−11.9	−0.75
Airborne									
SO₂ (lbs/ton shape)	12.52	−11.1	−1.57	10.36	−15.8	−1.95	14.01	−15.3	−2.54
Particulates (lbs/ton shape)	6.17	4.4	0.26	10.66	−11.6	−1.40	0.96	−13.5	−0.15
Land									
Solids (ton/ton shape)	0.12	−14.3	−0.02	0.18	−14.3	−0.03	0.04	−20.0	−0.01

Notes: Dashes = not applicable. Neg. = negligible.
[a] Raw ammonia liquor not used for coke quenching.
[b] Residuals modification equipment as in table 9.

for which the reductions are about 20 percent for the basic oxygen and electric arc, and about 35 percent for the open hearth.

The volatility of U.S. steel scrap prices is legendary. Ore prices, on the other hand, have been relatively stable.[19] Given the recent increased interest in recycling per se, it is important to assess how shifts in relative prices of ore and scrap would affect both scrap utilization and residuals. Table 12 illustrates the effects for the OH furnace. Note that substantial increases in ore prices relative to scrap prices have relatively little impact on the total quantity of scrap purchased, although there is a significant shift among types of scrap purchased. The limited response reflects the fact that product specifications limit the total amount of scrap that can be used because of the content of impurities. The major differences in residuals generation are the reduction in suspended solids and the increase in SO_2 generation. Increases in scrap prices result in much reduced use of scrap, with consequent increases in blast furnace capacity used and generation of BOD_5, phenols, ammonia, SO_2, and particulates.

Although the open hearth furnace, which is being decreasingly used in United States to produce steel, is used to illustrate the impacts of changes in raw material prices, analogous results, in terms of shifts in raw material utilization, would occur at more typical steel mill installations, for example, where combinations of the basic oxygen and electric arc types exist, or where 60/40 charge to the basic oxygen is practiced rather than 70/30, or both. Both of these configurations provide flexibility to respond to changing raw material prices. The point to be emphasized is that factors exogenous to the mill which affect raw material prices will have important repercussions on residuals generation at the mill.

Among the general conclusions to be drawn from the study of steel manufacture are the following:

1. The introduction of continuous casting on a broad scale will result in a significant decrease in the environmental quality problems created by the production of steel.

2. Although an increase in the price of scrap (caused for example, by growing export markets) may encourage the collection of scrap items that might otherwise end up as part of a diffuse litter problem, it will also discourage the use of scrap at domestic steel mills. The implication of

[19] This stability may well be a result of vertical integration in the industry with respect to ore as a raw material. Similar integration does not exist for steel scrap as a raw material.

TABLE 12.   Residuals Discharges at an Open Hearth Plant with Variations in Ore and Scrap Price Levels: Conventional Casting and Full Finishing[a]

Mill characteristics and residuals discharges	Bench mark case[b]	≥36% increase in ore prices[c]	100% increase in scrap prices[d]
Average steel cost ($/ton shape)	70.34	75.78	75.65
Marginal iron cost ($/ton hot metal)	40.61	43.40	41.79
Percent blast furnace capacity used	78.8	64.4	86.1
Scrap purchased (daily tons)			
#1 Factory	149.9	164.3	0.0
#1 Heavy melting	436.3	425.8	0.0
Shredded	0.0	70.0	0.0
#2 Bundle	165.6	93.8	211.9
Ore purchased (daily tons)			
High iron content	2,007.2	1,976.3	2,912.3
Low iron content	80.4	78.8	58.0
Residuals discharged[e]			
Waterborne			
BOD_5 (lbs/ton shape)	5.08	5.05	5.50
Oil (lbs/ton shape)	3.17	3.17	3.22
Phenols (lbs/ton shape)	0.52	0.51	0.65
Ammonia (lbs/ton shape)	1.21	1.19	1.50
Suspended solids (lbs/ton shape)	6.23	4.98	5.57
Heat (10^6 Btu/ton shape)	5.22	5.21	5.51
Airborne			
SO_2 (lbs/ton shape)	14.09	14.96	17.12
Particulates (lbs/ton shape)	5.91	5.99	7.34
Land			
Solids (tons)	0.14	0.14	0.18
Percent steel by			
70/30	1.1	0.0	84.8
50/50	98.9	100.0	15.2

a Raw ammonia liquor not used for coke quenching.

b 1968 price levels of ore and scrap. See table 9.

c Scrap prices at 1968 levels; increase in ore prices in relation to 1968 levels.

d Ore prices at 1968 levels; increase in scrap prices in relation to 1968 levels.

e Residuals modification equipment as in table 9.

smaller average scrap proportions in steel furnace charges is a larger liquid residuals problem for the steel industry.

3. In contrast, very low discharge levels of most waterborne residuals can be achieved without undue hardship by taking advantage of available modification and by-product recovery processes. Costs will be significantly higher if use of raw coke plant liquor for quenching is prohibited because of air pollution problems.

4. To the extent that a given plant has installed, for whatever reason, particulate control equipment with efficiencies comparable to those as-

sumed in the base case, large further reductions in particulate discharges will be quite expensive.

Automotive Steel Scrap

As the discussion in the preceding section indicated, steel scrap—and policies affecting its use—have important implications for residuals management in the steel industry. Therefore a separate study of one type of steel scrap, that produced from obsolete automobiles, was undertaken.[20] Although automobile scrap represented in 1970 only about 4 percent of total ferrous materials feed to steel production,[21] it is the most visible scrap problem because of the abandoned vehicles "discharged" onto the landscape.[22]

Figure 13 depicts the sequence of steps from automobile usage to production of scrap from obsolete automobiles. A linear programming model of the system—dismantling, transportation, and hulk processing operations—was developed. This is an example of the "system" type of industry study that involves several spatially separated processes and operations.

As emphasized previously, in terms of the impurities present in the steel, the manufacture of steel requires that certain physical property specifications must be met. These specifications are particularly critical for steels that are to be rolled into thin sheets and then formed into complex shapes, a process that requires keeping the content of alloying metals very low. The essence then of the problem of utilizing obsolete steel products, such as automobiles and appliances, is the amount of alloying materials in the obsolete product, the extent to which these materials can be removed in processing, and the related costs of doing so. Automobiles, for example, contain large amounts of nonferrous metals, primarily in the form of "add on" parts, such as wiring, generators, and radiators. If not removed in processing, most of the undesired materials (except chromium) are transferred from the scrap inputs to the steel output.

The study of automotive steel scrap focused on the technology and costs of dismantling and processing to produce different qualities of steel scrap. The dismantling model simulates the removal from an obsolete

[20] Sawyer, *Automotive Scrap Recycling*.

[21] Ibid., p. 14.

[22] See F. Lee Brown and A. O. Lebeck, *Cars, Cans, and Dumps: Solutions for Rural Residuals* (Baltimore, Johns Hopkins University Press for Resources for the Future, 1976).

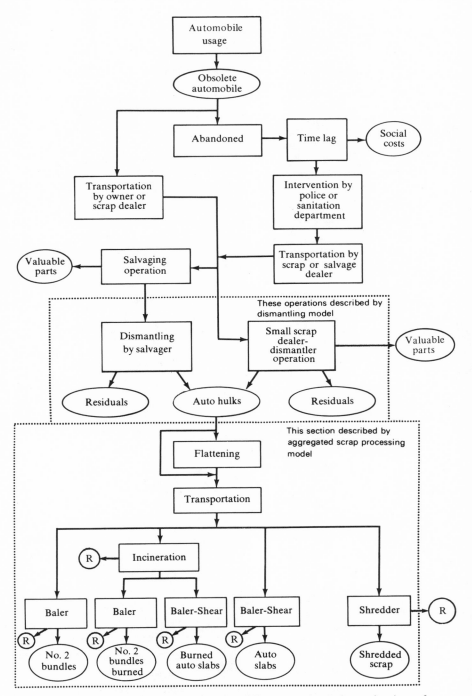

Figure 13. Sequence of steps from automobile usage through production of automotive steel scrap. From: Sawyer, *Automotive Scrap Recycling,* p. 112.

TABLE 13. Variable Inputs and Nonproduct Outputs (Residuals) Associated with Steel-Scrap-Processing Equipment

Factor	Type of process			
	Incineration	Baler	Shear	Shredder
Electricity (kWh /gross ton input)	3.00	10.20	6.40	42.5
Water (gal. /gr. ton proc.)	10.00	700.00	400.00	240.00
Gas (1,000 cu. ft. /car proc.)	0.08	0.00	0.00	0.00
Kerosene (gal. /car proc.)	7.00	0.00	0.00	0.00
Diesel fuel (gal. /car proc.)	0.00	0.00	0.00	1.75
Carb. monoxide (lbs. /gr. ton proc.)	3.61	0.00	0.00	0.00
Carb. dioxide (lbs. /gr. ton proc.)	3.61	0.00	0.00	0.00
Nitrogen oxides (lbs. /gr. ton proc.)	0.36	0.00	0.00	0.00
Particulates (lbs. /gr. ton proc.)	0.21	0.00	0.00	20.00
Capacity utilization as a fraction of capacity	0.90	0.90	0.90	0.90
SO_2 per car burned (lbs.)	0.90	0.00	0.00	0.00

Source: Sawyer, *Automotive Scrap Recycling*, p. 79.

TABLE 14. Quality of Scrap Output (Product) and Related Processing Costs[a]

Process (including hand dismantling as usually practiced)	Percent copper in in final product	Percent heterogeneous copper removed	Estimated total cost[b] of production in dollars per gross ton
Baling or shearing	0.50	50	18
Shredding	0.22	70	28
Improved shredding	0.12	80–85	30[c]
Cyrogenic shredding	0.08	90–95	35[c]

Source: Sawyer, *Automotive Scrap Recycling*, p. 111.

[a] Assumed hulk cost $10; assumed deregistered auto cost to dismantler $10; assumed hulk transportation cost $5; 1972 dollars.

[b] Cost exclusive of transportation of processed scrap and of profits.

[c] Very rough estimate—no data available.

automobile of up to forty parts, such as the radiator, engine, transmission, and fuel pump. For a given result of the hand dismantling operation, table 13 shows the inputs to, and residuals generated in, the main types of scrap processing. Table 14 shows the quality of scrap in terms of copper content, and the costs, for the processes other than incineration. Figure 14 depicts the cost–copper content relationship.

The results of the automotive scrap study amplify the results of the steel industry study as far as understanding the factors influencing the use of different types of steel scrap is concerned. The demand for scraps is of course a function of their prices. But these prices in turn are affected by various technological and economic factors associated with use of

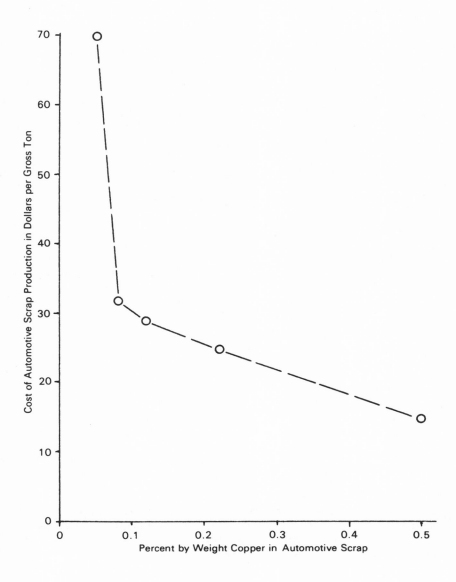

Figure 14. Estimated costs (exclusive of profit in 1972 dollars) of steel scrap produced from automobiles as a function of the amount of copper in the resulting scrap. Note: all plants are assumed to be sufficiently large so that further reductions in costs by increasing scale are negligible. Costs are exclusive of profit in 1972 dollars. From: Sawyer, *Automotive Scrap Recycling,* p. 112. 97

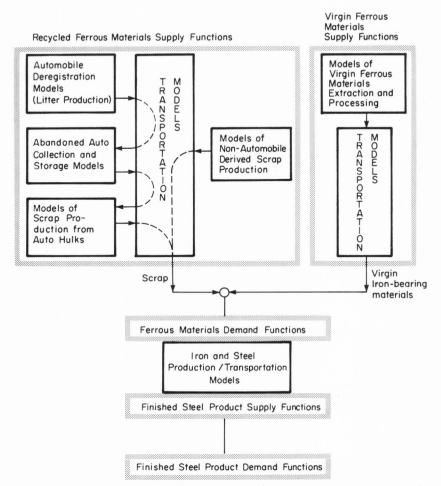

Figure 15. Total system structure for analysis of supply of steel. From: Clifford S. Russell and William J. Vaughn, *Steel Production: Processes, Products, and Residuals* (Baltimore, Johns Hopkins University Press for Resources for the Future) p. 290.

scrap and other ferrous raw materials and the characteristics of the steel and steel products produced. These factors include: (1) the mix of furnace types used in the steel industry, since the open hearth, the electric furnace, and the basic oxygen processes have different capacities for using scrap; (2) the product mix of steel production, that is, if a larger fraction of the steel produced had less severe constraints on nonferrous content, the quality constraints on use of steel scrap would be less restrictive; (3) the price of prereduced ore, which serves as a constraint on the maximum level scrap prices can reach, since they are substitutable goods, even in an electric furnace; (4) the design of motor vehicles, and other goods, in terms of the proportion of nonferrous materials included and their forms, whether they are separable or inextricably mixed; and (5) the costs of residuals management in both scrap processing and steel production. What is needed then is an analysis of the total system for producing steel, with: (a) various degrees of recycling; (b) various combinations of steel production processes; (c) various degrees of scrap processing; (d) various designs of products using steel, and thus various mixes of steel outputs; and (e) costs of required residuals management at all points in the system. Such an analysis would allow the explicit determination of the effects of alternative economic policies and technological constraints on the use of automotive (and other) scrap and the extent of recycling. Figure 15 illustrates the framework of this type of analysis. Examples of the types of policies which could then be investigated include: [23]

policies affecting the supply of recycled ferrous materials
    subsidy for scrap collection, that is, of abandoned automobiles
    subsidy on scrap-processing equipment
    inventory tax on accumulated obsolete automobiles held by auto
        wreckers
    disposal fee on deregistered automobiles
    revised transportation rates for various scraps
    subsidy to scrap exporters
    restrictions on scrap exports
    restrictions on scrap imports

policies affecting the supply of virgin ferrous materials
    elimination or reduction of the iron ore depletion allowance
    revised iron ore transportation rates
    tighter restrictions on residuals discharges in mining and ore-
        processing operations

policies affecting ferrous materials demand
    subsidy on automobile-derived scrap purchases
    subsidy for investment in scrap-intensive steel production technol-

[23] Russell and Vaughan, *Steel Production,* chapter 12.

ogies (electric arc furnace or scrap presmelting at basic oxygen facilities)

policies affecting finished steel product demand and design
   tax on automobile manufacturers for use of copper and other key
      tramps in major components such as wiring, and the like
   regulation of automobile design standards

The Coal-Electric Energy Industry

Like the automotive scrap study, the focus of the coal-electric (CEEI) energy study is a set of spatially separated operations to produce a final product: coal mining; coal preparation, which generally takes place at or very near the mine; coal or energy transport—the latter being relevant when energy is generated at the mine; and energy generation via coal combustion. Figure 16 depicts the interrelated elements of the system, the types of residuals generated at each of the operations, and some of the posssible residuals modification alternatives. The major residuals involved are:

   acid mine drainage from underground mining
   overburden from strip and open pit mining (surface mining)
   suspended solids in coal preparation plant wash water
   particulates from air-flow cleaners and thermal driers at coal prepara-
      tion plants
   particulates in power plant gaseous discharges
   sulfur oxides in power plant gaseous discharges
   waterborne heat from the power plant

The problem addressed is as follows: assuming a given energy demand at the high-voltage side of the load center substation, what is the minimum-cost system of coal production, coal processing, coal or energy transport, coal use in the power plant, and residuals-handling activities that will meet specified discharge standards applying to liquid and gaseous residuals and the disposition of solid residuals at all points in the system? Tradeoffs are possible in such a system, for example, among raw coal quality, degree of coal preparation, transport of coal or energy, combustion technology, and residuals-handling technology at the mine-prepara-

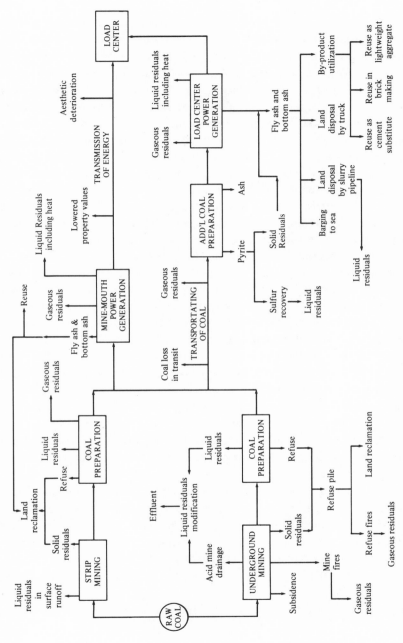

Figure 16. Residuals generation and some residuals modification alternatives in the use of coal to produce electric energy

tion plant and the power plant. If the objective is to reduce the discharge of SO_2, various combinations are possible: using high or low sulfur coal, various degrees of coal preparation, and various options for removal of SO_2 in the stack of the power plant. Thus at least a partial alternative to removal of residuals in the stack gas is to increase the degree of coal preparation for the removal of sulfur (and ash). However, although doing this reduces the residuals generated at the power plant, it increases the residuals generated in coal mining and in coal preparation, because more hydrocarbons are discarded, thereby requiring more coal to be mined for the same Btu input to the power plant. Modifying power plant flue gas and use of cooling towers to reduce thermal discharges to watercourses are both energy-intensive measures, thus requiring more coal to be mined to produce the same net energy output and hence generating more residuals in coal mining and coal preparation. A qualitative indication of these types of interrelationships is presented in table 15.

A strategy to reduce the discharge of one residual may add to the costs of handling another residual or generate additional residuals which result in other residuals–environmental quality problems. Removing sulfur in power plant flue gas increases the resistivity of the fly ash, thereby making electrostatic precipitation less efficient. The use of cooling towers to reduce thermal discharges into watercourses involves a transfer of the residual heat from a liquid discharge to a gaseous discharge. This discharge from a cooling tower may result in such undesirable environmental effects as local fogging and icing, cloud formation, and increased precipitation.[24] In turn, various alternatives are available for modifying these effects, such as superheating the plume or using finned heat exchangers.[25] Any such modification adds to residuals modification costs, and to the energy required.

In considering energy generation systems, that is, sets of operations and processes used to transform potential energy in place—coal, petroleum, natural gas, uranium—to usable energy in terms of kilowatt hours or British thermal units, it is essential to include in the analysis all energy inputs required throughout the system to produce the desired output, including energy used in residuals modification and disposal. Thus, in the

[24] See Eric Aynsley, "Cooling Tower Effects: Studies Abound," *Electrical World* vol. 173, no. 19 (1970) pp. 42–43.

[25] Hennie Veldhuizen and Joe Ledbetter, "Cooling Tower Fog: Control and Abatement," *Journal of the Air Pollution Control Association* vol. 21, no. 1 (1971) pp. 21–24.

TABLE 15.   Effects of Strategies for Improving Environmental Quality in the Utilization of Coal to Produce Electric Energy
(0 = no change; − = less; + = more)

	Strategy[a]									
	1	2	3	4	5	6	7	8	9	10
Effects on system										
Quantity of coal mined	0[b]	0	0	0	−[b]	+[b]	−	+	0	+
Quantity of raw mine drainage	−	0	0	0	−	+	−	+	0	+
Quantity of refuse at mine	0	+	+	+	−	+	−	+	0	+
Quantity of coal transported	0	0	0	0	−	−	−	+	0	+
Quantity of solid residuals at power plant	0	0	0	0	−	−	−	+	0	+
Effects on residuals										
Useful land	+	0	0	0	+	−	+	−	0	−
Acid and iron generated	0	−	0	0	−	+	−	+	0	+
Suspended solids generated at preparation plant	0	0	−	0	−	+	−	+	0	+
Particulates generated at preparation plant	0	0	0	−	−	+	−	+	0	+
Particulates and sulfur oxides generated at power plant	0	0	0	0	−	−	−	−	0	+
Residual heat generated at power plant	0	0	0	0	−	0	0	+	0	−
Suspended solids generated at power plant	0	0	0	0	−	−	−	+	−	+
Solid residuals generated at power plant	0	0	0	0	−	−	−	+	+	+

[a] *Strategies:*
  1. Grade and replant land
  2. Treat acid mine drainage
  3. Treat waste water from preparation plant
  4. Collect particulates from preparation plant
  5. Increase generator efficiency
  6. Use more coal preparation
  7. Use higher quality *raw* coal
  8. Treat power plant flue gas
  9. Treat suspended solids from power plant
  10. Use cooling towers
[b] All + changes represent negative effects on environmental quality, except for land.

CEEI system, energy is required to mine and transport the coal, to replace or otherwise handle the overburden, and to reestablish vegetation on the land disturbed by mining.

As in the case of the pulp and paper study, the method of analysis in the CEEI study was finite sampling of the response surface to determine minimum total costs—production costs plus residuals management costs —to produce a specified output, a net of 10 billion kilowatt hours per year at the high side of the busbar at the load center. Two cases were analyzed, both of which assumed a base load plant of two units, each of which had net generating capability of 800 megawatts and an assumed average lifetime plant factor of 71 percent.[26] All units in the system are sized to provide the inputs needed to produce the specified output.

Case A is a system comprised of: a mine-mouth power plant; an area-strip mine with 80 feet of overburden, 4-foot coal seam, and high sulfur, high ash, high moisture, low Btu content coal, and 400 mile transmission distance to the load center substation. Case B is a system with: a load center power plant; a 400-foot underground mine, 4-foot coal seam, and low sulfur, low ash, low moisture, high Btu content coal; unit train transport of coal for 400 miles; and 25-mile transmission distance from power plant to load center substation. The specific characteristics of the coals used are:

Case A (high ash, high sulfur, strip-mined coal)

High-volatile C bituminous coal, Illinois No. 6 seam; calculated calorific value (dry basis) is $[100 - (A + S)] H + 40.5S$, where A is % ash, S is % sulfur, and the "H value" for this type of coal is about 14,350 Btu per lb. The ash-softening temperature for the various size and gravity fractions range from 1,980 to 2,140°F. Face-sample, as-received basis, contents: 14.0% moisture, 21.5% ash, 29.0% volatile matter, and 35.5% fixed carbon. Sulfur content was 3.8% and the calorific value 8,870 Btu per lb. Size analysis: +⅜″, 66.2%; ⅜″ × 14 meters, 28.0%; 14 meters × 100 meters, 4.3%; 100 meters × 0, 1.5%.

Case B (low ash, low sulfur, underground-mined coal)

High-volatile A bituminous coal, No. 2 Gas seam; calorific content calculated from an H-value of 15,460 Btu/lb (dry basis). The ash-softening temperature is 2,420°F for the prepared coal; in the calculations, data were converted from a face-sample, as-received basis, containing 1.2% moisture, 6.2% ash, 36.1% volatile matter, and 56.5% fixed carbon. The sulfur content was 0.9% and the calorific

[26] Initially such a plant would be operated around the clock, except for regularly scheduled maintenance. The plant factor typically decreases with increasing age of the plant, until it is retired from service. The analyses could have been made to reflect the higher plant factor in the early years and the "tapering off" process as the plant ages. However, the calculations showed that such a refinement would have little impact on the choice of residuals management options.

value 14,190 Btu per lb. Size analysis: + ⅜″, 64.6%; ⅜″ × 14M, 28.3%; 14M × 100M, 5.9%; 100M × 0, 1.2%. The washability characteristics of the 1½″ × 100M face-sample (dry basis) were, at 1.58 specific gravity, yield 95.6%, ash 3.3%, sulfur, 0.76%; at 1.45 specific gravity, yield 93.7%, ash 3.0%, sulfur 0.75%; at 1.35 specific gravity, yield 89.1%, ash 2.5%, sulfur 0.74%. When the sample was crushed to ⅜″ top size, there was no change in the washability characteristics.

Five coal preparation processes were posited, reflecting increasing degrees of preparation:

P_1 crushing to 1.5 inch top size—no cleaning
P_2 wet cleaning of coarse size coal, blended with uncleaned fines
P_3 wet cleaning of coarse size, airflow cleaning of fines
P_4 wet cleaning of coarse and fine sizes
P_5 wet cleaning of coarse and fine sizes at low specific gravity

The extent to which water used in coal preparation will be recirculated is a function of the cost of intake water versus the cost of in-plant water recirculation. In coal preparation processes, such as P_4 and P_5, where all coal particles including the very fine sizes are processed, part of the material suspended in the wash water is often recovered as usable coal; the suspended particles with a high-ash content would be disposed of as solid residuals. Increasingly the value of the coal recovered compensates for the cost of installing and operating flotation cells and vacuum filters; however, the cost of handling water that contains the particles remains.

Three sets of increasingly stringent constraints on residuals discharges were imposed on each system, characterized as "minimal," "moderate," and "high" (strict), and referred to as Environmental Quality Levels I, II, and III, respectively. Level I represents a minimal level of restriction on residuals discharges, consisting basically of those that would be undertaken for economic reasons, plus—for gaseous residuals—the level of particulate reduction typical prior to the Clean Air Act of 1967. Level III represents approximately the level of presently available or near-future technology, plus a high level of reclamation of land disturbed in strip mining. Level II is intermediate. The three levels are given more specificity in tables 16, 17, and 18. Tables 16 and 17 show for the two cases (1) the effects of the different degrees of coal preparation on the quality of the delivered coal and (2) the residuals discharges associated with coal

TABLE 16.   Degree of Coal Preparation Versus Quality of Delivered Coal

Preparation process	Case A: High-impurity seam mineral impurities (thousand tons/year)		Case B: Low-impurity seam mineral impurities (thousand tons/year)	
	Ash	Sulfur	Ash	Sulfur
1	1,210	210	340	32
2	870	170	230	28
3	520	140	190	28
4	350	120	180	24
5	230	110	80	24

TABLE 17.   Residuals Discharges Versus Degree of Preparation and Level of Environmental Quality[a]

Case A. High-impurity seam

Preparation process	Mining residual land (acres/year)			Preparation plant residuals				
	Disturbed	Regraded		Water, 1–8% solids (million gallons/year)		Airborne dust (tons/year)		
		I	II & III	I	II & III	I	II	III
1	980	0	980	0	0	0	0	0
2	1,014	0	1,014	364	0	0	0	0
3	1,114	0	1,114	400	0	1,200	4	4
4	1,042	0	1,042	2,780	0	70,000	4,400	860
5	1,250	0	1,250	2,628	0	70,000	4,400	860

Case B. Low-impurity seam

Mining residual, acid drainage	Preparation process	Gross increase in mine drainage (million gal/yr)[b]			Gross increase in acid content (tons H_2SO_4/yr)		
		I	II	III	I	II	III
	1, 2, 4	110	80	80	360	180	0
	3, 5	120	90	90	400	200	0
Preparation plant residuals	approximately 60% of those for Case A						

[a] For 1,600 net MW output at 71% load factor.
[b] Average flow rate for underground mines in Pennsylvania.

mining and coal processing under the three environmental quality levels. Table 18 summarizes the constraints on the discharges for the two cases and the three levels.

Table 19 shows the results of the analyses of the two cases for the three environmental quality levels. For Case A, the least-cost system utilized coal preparation processes P_1, P_4, and P_5 for levels I, II, and III, respectively; for Case B, P_1, P_1, and P_5 for the three levels, respectively. The relative costs, compared with Level I, are:

	I	II	III
		(dollars)	
Case A	1.00	1.18	1.54
Case B	1.00	1.14	1.39

The incremental costs are distributed approximately as follows for Case A: From Level I to Level II—about one-third to coal mining and preparation, both production and residuals management costs; about two-thirds to the power plant for a higher level of gaseous residuals modification and the shift from once-through cooling to a wet cooling tower, plus compensating generating capacity. From Level II to Level III—primarily to the power plant, where over 90 percent is attributable to the shift to a dry cooling tower, and secondarily to the more expensive design of the transmission lines. Less than 5 percent of the increased cost is attributable to the higher level of modification of the gaseous residuals. For Case B: from Level I to Level II—virtually all of the increase to the power plant, attributable to the higher level of gaseous residuals modification and to the shift to a wet cooling tower, plus compensating generating capacity; from Level II to Level III—as with the mine-mouth plant, primarily to the power plant and transmission.

It should be emphasized that the results relate to new facilities throughout the system and to a *single* power plant abstracted from a total power system. An actual power system would consist of multiple types and locations of energy-generating units—base load and peak load, multiple transmission lines, interconnections with adjacent power systems, and multiple sources of coal (and other energy sources). These different units would typically be of different ages with different efficiencies, both with respect to heat rates and residuals generation. There would be additional options for achieving specified levels of *ambient* environmental quality (rather than limiting residuals discharges, per se). Examples would be load shifting[27] and purchase of energy. Nevertheless, even the analysis of this simplified coal-electric energy system illustrates both the range of residuals management options available and the relative magnitude of the effects of restrictions on residuals discharges on energy costs. The

[27] Donald S. Shepard, "A Load-Shifting Model for Air Pollution Control in the Electric Power Industry," *Journal of the Air Pollution Control Association* vol. 20, no. 11 (1970) pp. 756–61.

TABLE 18. Characterization of Environmental Quality Levels: Cases A and B

	Environmental Quality Level		
	I–Minimal	*II–Moderate*	*III–High (strict)*
		Case A. Mine-mouth power plant	
Coal mine (area strip)	No grading; minimum erosion control	Grading; high degree of erosion control	Same as for II, but with additional reclamation, the benefits of which offset the costs
	Cost: $100/acre	Cost: $500/acre	Net cost: $500/acre
Coal preparation plant	No control	Closed circuit water system; scrubbers for particulate control	Closed circuit water system; scrubbers and air recirculation for particulate control
Power plant[a] Flue gas equipment	Mechanical collector for particulates; no SO_x, NO_x reduction	Wet scrubber with pulverized limestone added to scrub H_2O; flue gas reheat for NO_x reduction	Two wet scrubbers in tandem, using limestone, or wet scrubber followed by wet-type electrostatic precipitator
Stack height	300 feet	900 feet	500 feet
Cooling system	Once-through condenser	Evaporative cooling tower	Dry cooling tower
Solid residuals disposal	Bottom ashes sluiced to settling basin;	Bottom ashes—same as for Level I	Bottom ashes—same as for Level I
	Flyash conveyed pneumatically to storage silo	Flyash plus limestone reaction products pumped as a slurry, 10% solids, for ¼ mile to storage pond	Flyash—same as for Level II
Energy transmission	Standard lattice-type transmission towers throughout; no underground line	About 10% longer route than for Level I; about 45% tubular pole-type transmission towers; no underground line	100% tubular pole-type transmission towers; 10% of line underground

Case B. Load center power plant

Coal mine (underground)	No control	Prevention of contact between groundwater and oxidized pyritic material	Prevention if possible; neutralization of acidic mine drainage if formed
Coal preparation plant	Same as Case A	Same as Case A	Same as Case A
Coal transportation—unit train	Negligible residuals generated	Negligible residuals generated	Negligible residuals generated
Power plant[a] Flue gas equipment	Electrostatic precipitator with 90% particulate collection efficiency; no SO_x, NO_x reduction	Same as Case A	Same as Case A
Stack height	500 feet	900 feet	500 feet
Cooling system	Same as Case A	Same as Case A	Same as Case A
Solid residuals disposal	No control on scrubber effluent	Scrubber effluent settled in gravitational thickener, followed by sludge dewatering with vacuum filter; filter cake hauled to land disposal	Scrubber effluent—same as Level II
Bottom ashes	Same as Case A	Same as Case A	Same as Case A

[a] Two 800 net megawatt units, dry bottom-furnace.

TABLE 19. Cost of Electric Energy and Residuals Discharges for Three Levels of Environmental Quality: Coal-Electric Energy System

	Case A Mine-mouth plant, high-impurity seam, area-strip mine			Case B Load-center plant,[a] low-impurity seam, deep mine		
	Level I	Level II	Level III	Level I	Level II	Level III
Cost of energy (mills/kWh)						
At busbar	6.00	7.05	8.57	6.94	7.92	9.37
At load-center substation	7.64	9.00	11.78	7.17	8.20	9.98
Residuals discharges	(Annual basis, except for transmission)					
Acres of disturbed land/acres of reclaimed land	[980/0]	[1040/1040]	[1250/1250]	rs	rs	rs
Gross increase in mine drainage						
Million gallons	rs	rs	rs	110	80	90
Tons of sulfuric acid	rs	rs	rs	360	180	0
Preparation plant water (1 to 8% solids) (million gallons)	0	0	0	0	0	0
Preparation plant solid residuals (tons)	0	1,600,000	2,600,000	0	0	300,000
Preparation plant particulates (tons)	0	4,400	900	0	0	700
Power-plant stack discharges (tons)						
Particulates	140,000	900	100	25,000	700	100
Sulfur oxides (sulfur content)	200,000	29,000	15,000	28,000	7,500	3,500
Nitrogen oxides (nitrogen content)	10,000	8,000	2,000	10,000	8,000	2,000
Power-plant solid residuals (tons)	1,000,000	950,000	800,000	300,000	450,000	300,000
Thermal discharge to watercourse (billion Btu)	6,000	0	0	6,000	0	0
Water consumption (extra evaporation) (million gallons)	7,500	5,000	0	7,500	5,000	0
Transmission						
Lattice-type line towers (number)	4,000	2,500	0	120	0	0
Tubular pole-type line towers (number)	0	2,000	4,000	0	150	0
Underground circuit-miles/total circuit-miles	[0/800]	[0/880]	[80/800]	[0/25]	[0/28]	[25/25]

Notes: Two 800 net MW units, 71% average lifetime load factor. Costs are based on: 1968 prices; 8% return on power plant investment; 10% on coal mining/preparation investment. Power plant costs include 27% overhead covering architectural/engineering fees, construction equipment, and price escalation during construction. rs = relatively small.

110

same approach is relevant to assessing the environmental quality problems and residuals management costs of other types of fuel–energy generating systems.

Concluding Observations

The Quality of the Environment program industry studies have resulted in numerous insights about the utility, findings, problems, and limitations of such studies. We summarize a few of these.

Utility and Findings of Industry and Activity Modeling

1. The studies provide insights into the ways of responding to different REQM strategies and their approximate costs. This is of use in the typical REQM context (metropolitan area, region, or state) in which decisions are made about the type and stringency of direct or indirect constraints to be imposed on residuals discharges or dischargers.[28] The same information is essential for interregional analyses of the effects of REQM policies on the regional distribution of production, as in the studies by Heady and coauthors.[29] Similarly, the information is essential for macroanalyses seeking information on aggregate national costs among alternative REQM strategies to achieve the same ambient environmental quality goals. Even though the level of sophistication or degree of refinement of industry models for these uses will necessarily be substantially less than in an individual industry study, some detailed study is essential to identify critical variables. These variables will serve as a basis for developing the more simplified models that are to be components of regional, interregional, and macroanalyses. (Note that one such regional model is that of the Delaware Estuary described in chapter 7.)

2. One of the most important lessons from the industry studies is the important effect that exogenous factors can have on residuals generation and REQM costs in industry, even when such factors superficially appear to be unrelated to residuals. Examples include: (a) product specification

[28] The problems of modifying the detailed industry models for use in regional analysis are discussed by Russell in *Residuals Management in Industry*, Ch. 1 and pp. 181–85.

[29] Earl O. Heady and coauthors, *Agricultural and Water Policies and the Environment: An Analysis of National Alternatives in Natural Resource Use, Food Supply Capacity and Environmental Quality* (Ames, Iowa, Center for Agricultural and Rural Development, Iowa State University, June 1972) (CARD Report 40T).

changes; (b) shifts in relative prices of factor inputs, such as alternative raw materials (primary versus secondary), energy, water; and (c) induced technological change in basic production processes, and in materials recovery, by-product production, and residuals-processing to recover materials. These influences have typically been overlooked in past and current "pollution control" legislation and implementation and in the development of REQM strategies for specific regions.

3. There are many possible responses by plant management to constraints—that is, standards or charges—imposed on the discharge of residuals to the various environmental media. Such responses include changing raw material, production process, and product output specifications, plus materials recovery, by-product production, and conventional residuals modification. When a plant has many production processes—such as an integrated paper mill or a steel mill—and is producing multiple products, there is substantially greater flexibility to respond to changes in relative prices of factor inputs and to effluent controls.

That changing product specifications and changing product mix are options which have been adopted is clear from industry behavior in the last several years. For example, Georgia-Pacific Corporation announced in December 1973 that all of its paper towel production for institutional and industrial consumers would be converted to unbleached cellulose from the traditional white.[30] Reasons cited included reduced costs of residuals management, less energy required, and more possibility of using paper residuals as raw material. Other examples have been recorded, particularly in response to the recent sharp increases in energy costs.[31]

4. Even for major residuals-generating industries, analyses show that, up to relatively high levels of reduction in residuals discharges per unit of product or per unit of raw product processed, the proportion of total production costs represented by residuals management costs amounts to only a few percentage points. However, as "zero discharge"[32] of liquid and

[30] *Official Board Markets,* December 29, 1973, p. 1.

[31] See "The Squeeze on Product Mix," *Business Week* no. 2312 (January 5, 1974) pp. 50–55; and "Energy and the Product Mix," *Chemical Week* vol. 115, no. 22 (November 27, 1974) pp. 24–32.

[32] Zero discharge of all residuals is of course impossible, even neglecting the residuals normally considered innocuous, that is, CO_2, water vapor, heat to the atmosphere, and dissolved solids. The inevitable consequence of a zero discharge policy for liquid and gaseous residuals is an increase in the quantity of solid residuals requiring disposal.

gaseous residuals is approached, the marginal costs of reduction in discharges increases substantially, and residuals management costs may become an important proportion of total production costs. Economies of scale, multiproduction processes, and multiproduct outputs reduce the per unit costs of residuals management.

5. There are physical, technological, and economic linkages among liquid, gaseous, solid, and energy residuals. Often management strategies directed toward a single type of residual have overlooked such linkages, in terms of the effects a response to reduce the discharge of one type of residual may have on other residuals simultaneously and in terms of the generation of secondary and tertiary residuals, and on the additional inputs required for reduction in the discharge of primary residuals, especially energy. The industry studies described here explicitly considered these effects and demonstrated the necessity for developing integrated residuals management strategies rather than separate strategies. (Similar demonstration resulted from the Delaware case study on the regional level. This case study is reported in chapter 7.)

6. Although most of the analyses in the RFF industry studies assumed grass roots rather than existing plants (the study of the steel industry being the major exception), the results in terms of what factors influence both residuals generation and the responses to effluent controls are relevant to both types of plants. The major difference between existing and "grass roots" plants involves the constraints imposed by the physical layout and location of the existing plant. With respect to both physical possibility and absolute costs, many residuals management options are site and plant specific, the latter because of existing plant equipment and layout, product mix, type of plant construction, and site characteristics (including climate). These constraints may shift the sequence of options adopted at a plant or preclude certain options because of excessive costs. But the direction of response is the same for both existing and new plants.

This point has another important implication. Because of the inability to consider site-specific options, many, if not all, estimates of REQM costs aggregated to the national level for an industry depend on cost functions for conventional end-of-pipe options for residuals modification. This results in near-maximum costs to achieve any given level of discharge reduction (for a given set of factor input prices) since the estimates neglect the rich menu of responses that are usually possible in actual situations.

7. The factors that affect residuals generation in industry and REQM strategies both directly influence the use of secondary materials, such as steel scrap and paper residuals. This is explicitly demonstrated by the steel, automotive scrap, and pulp and paper studies. This means that policies expressly directed toward increasing the use of secondary materials may or may not be successful and may exacerbate rather than reduce REQM problems.

8. In many cases, the relevant focus for analysis of residuals management is a system of spatially separated but linked activities to produce a final product or set of products, such as those described in the coal-electric energy study. What is needed are industry models that include the totality of processes and operations associated with a given product, from raw material extraction through use of the product and disposal after use, and the corresponding inputs and residuals generation with their management-discharge options. This focus is shown in figure 15 and in figure 17. The latter shows alternative total systems associated with a given paper product. One important next step in the analysis of residuals management in industry is to generate studies with this expanded scope.

Problems and Limitations of Industry Studies

1. Whatever the method of analysis used in an industry study—programming or sampling—basic information is needed about material and energy balances involved in the activity and cost functions of alternative methods of reducing the discharge of residuals. Developing such balances and cost functions requires extensive and detailed knowledge of the technology of a given industry or activity, at least where something more than a "broad brush" analysis is involved. Even where a single existing industrial plant is the subject of the analysis, the full range of possibilities for residuals modification can rarely be assessed, even when the type of raw material, production process, product mix, and product output specifications are all held constant. Because the outputs are limited by the inputs, the results reflect only the options included in the model. Nevertheless, the experience at RFF shows that a rich range of options can be incorporated in the industry models.

2. The models developed to date do not reflect the variability in residuals generation that exists in the real world. To what extent residuals management costs in industry (and other activities) in the real world reflect the necessity of dealing with that variability has not been analyzed.

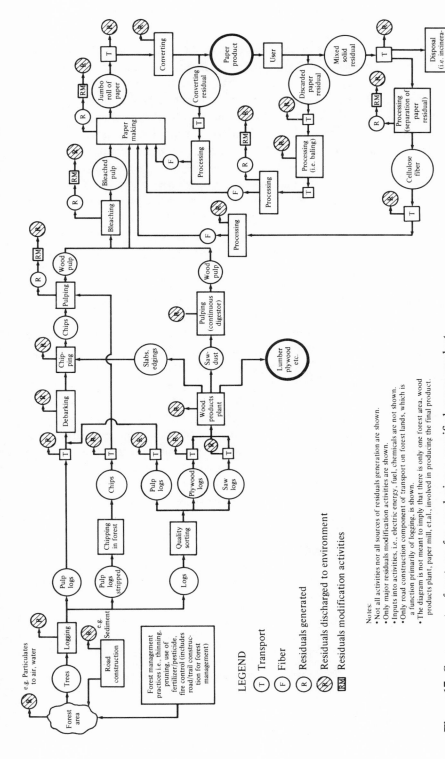

Figure 17. Components of systems for producing a specified paper product

LEGEND

T Transport

F Fiber

R Residuals generated

⚡ Residuals discharged to environment

RM Residuals modification activities

Notes:
- Not all activities nor all sources of residuals generation are shown.
- Only major residuals modification activities are shown.
- Inputs into activities, i.e., electric energy, fuel, chemicals are not shown.
- Only road construction component of transport on forest lands, which is a function primarily of logging, is shown.
- The diagram is not meant to imply that there is only one forest area, wood products plant, paper mill, et.al., involved in producing the final product.

3. All modeling is done in a world in flux. This means that often by the time the model is built, prices of factors of production and final products have changed, sometimes substantially. This in turn may lead to significant changes in product mix, raw materials, and product specifications even at an existing plant. For a new plant, such changes in prices are likely to lead to selection of different technology of production.

The dynamic context of activity modeling puts a premium on models that are flexible. The model should contain both a reasonably complete universe of possible residuals modification alternatives and a universe of other possible responses to changing factor prices, at least those which can be foreseen as being likely to change. It is about as difficult to predict changes in prices for such inputs as energy, as it is to predict changes in certain variables such as quality of raw material.[33] What the models need to demonstrate is what responses are possible and likely in reaction to changes in factor prices.

4. The costs directly attributable to constraints on residuals discharges are not always easy to determine. Theoretically, the baseline for developing REQM costs should be the basic production, materials recovery, and by-product activities that would take place in the absence of controls on residuals discharges. This may or may not be easy to determine, even with detailed analysis. A related problem is that some options to reduce residuals discharges involve changes in a number of factor inputs—materials, energy, equipment—which changes may be increases or decreases, such as in the replacement of a direct contact evaporator with an indirect contact evaporator or the building of excess capacity in a recovery furnace in a kraft mill to reduce residuals generation or discharge. Thus, to determine the cost associated with reducing residuals discharge requires analysis of alternative *complete* sets of processes and operations and their respective inputs and outputs.

5. Finally, it should be emphasized that, although the discussion in this chapter has used manufacturing activities for its examples, the same conceptual approach is relevant for other residuals-generating activities—agricultural, silvicultural, mining, commercial, residential, and institutional.

[33] For example, as a function of genetic changes through fruit and vegetable development, or changes in product specifications over time, such as increasing brightness of paper products or increasing the strength of steel at the same time thinner steel is produced.

Appendix 4-A
Use of Linear Programming for Analysis of
Integrated Residuals Management at Industrial Plants

We mentioned in the body of this chapter that linear programming offers a particularly powerful analytical tool for the analysis of REQM problems at the individual activity level. There are two reasons for this: (1) it provides an explicit and efficient optimum-seeking procedure; and (2) once formulated, results from changing variables of interest can be calculated very rapidly. This appendix sketches a linear programming model used in the petroleum refining and steel industry studies discussed in the chapter proper. Familiarity with basic linear programming is assumed.

A first problem, of course, is that the functions involved are inherently nonlinear. Thus if one wishes to capture them fully, the resulting model is complex and difficult to solve even for simple situations. To avoid this problem, the approach used in constructing working models was to attempt to identify only some relatively small set of discrete production alternatives and to structure these in the form of unit activity vectors, that is, vectors giving inputs and costs required for the processing of one unit of an input or production of one unit of output of interest. The objective of the firm may be taken to be profit maximization, cost minimization for given output, or any other convenient variant. The constraint set may include limits on input availability; product mix; quality requirements to be met by products; limits on discharges of one or more residuals; and, most important, continuity conditions (or mass and energy balance equations) requiring, for example, that the full amount of each residual generated be accounted for explicitly, either by intake to a modification or transport process or by discharge.

To say that once a quantity of residual is generated in the model, it must be modified, transported, or discharged is not to say that *all* residuals must necessarily be included in the model. For many purposes, for example, the carbon dioxide and water residuals from combustion processes will not be of interest and may be ignored in the construction of a response model. Which residuals are of interest will depend to a large extent on the spatial and time dimensions of the study, on the particular REQM problem being investigated, or both.

AUTHORS' NOTE: This appendix is adapted from Clifford S. Russell, *Residuals Management in Industry: A Case Study of Petroleum Refining* (Baltimore, Johns Hopkins University Press for Resources for the Future, 1973).

Columns / Rows	Production alternatives X_1 $\cdots$ X_H	By-product production B_1 $\cdots$ B_J	Raw material recovery W_1 $\cdots$ W_K
Production and sale	$+\bar{e}x_1$ $\cdots$ $+\bar{e}x_H$	$+\bar{b}_1$ $\cdots$ $+\bar{b}_J$	
Input availability	$-\bar{p}x_1$ $\cdots$ $-\bar{p}x_H$		$+\bar{w}_1$ $\cdots$ $+\bar{w}_K$
Output quality	$+\bar{q}x_1$ $\cdots$ $+\bar{q}x_H$		
Primary residuals	$-\bar{r}x_1$ $\cdots$ $-\bar{r}x_H$	$+\bar{e}_{B_1}$ $\cdots$ $+\bar{e}_{B_J}$	$+\bar{e}_{W_1}$ $\cdots$ $+\bar{e}_{W_K}$
Secondary residuals		$-\bar{r}_{B_1}$ $\cdots$ $-\bar{r}_{B_J}$	$-\bar{r}_{W_1}$ $\cdots$ $-\bar{r}_{W_K}$
Possible discharge constraints			
Objective function	Costs of production	Costs of production	Costs of recovery

Figure 4-A-1. Schematic of models of industrial residuals management. Note: The $\bar{e}$ are column vectors of zeros and ones. A particular vector $\bar{e}$ has the number of row elements corresponding to the constraint set in which it appears. The occurrence of ones is determined by the function of the column

The term "secondary residuals" is used to refer to residuals generated in such auxiliary processes as recirculation (including materials, energy recovery, or both), by-product production, and residuals modification, in contrast to those generated in the production process itself (the "primary residuals"). This distinction is made for convenience only, for as we describe below, secondary residuals are subject to the same mass-energy balance (continuity) conditions as are primary residuals.

The general form of the model is shown schematically in figure 4-A-1 and may be interpreted as a familiar linear programming problem

maximize $c'x$
subject to $Ax \leq b$
and $x \geq 0$

The activity levels x which are to be chosen by the solution process are shown across the top under six headings: production alternatives (X_1,

Treatment and transport of residuals $T_1 \;\cdots\; T_L, \quad V_1 \cdots V_M$		Sale of products $Y_1 \;\cdots\; Y_N$	Discharges of residuals $D_1 \cdots D_g \;\cdots\; D_G$	Right-hand side
		$-\bar{e}_{Y_1} \cdots -\bar{e}_{Y_N}$		$\geq \quad 0$
				$\geq \quad -\tilde{P}$
				$\genfrac{}{}{0pt}{}{\geq}{\leq} \quad \tilde{Q}$
$+\bar{e}_{T_1}$	$+\bar{e}_{V_1} \cdots$		$\cdots +\bar{e}_{D_1}$	$= \quad 0$
$-\tilde{r}_{T_1} \cdots +\bar{e}_{T_L} \cdots +\bar{e}_{V_M}$ $\qquad -\tilde{r}_{V_1}$ $\qquad\qquad -\tilde{r}_{V_M}$ $\qquad -\tilde{r}_{T_L}$			$+\bar{e}_{D_g}$.	$= \quad 0$ . .
Etc.			. $+\bar{e}_{D_G}$	$= \quad 0$
		$+1$		
		$+1$		$\leq \quad \tilde{F}$
			$+1$	
Costs of treatment and transport		Prices of output	Possible effluent charges	

in which the vector appears. Thus in $\bar{e}_{x1}$, a one appears in the row corresponding to the output of process X_1. From: Russell, *Residuals Management in Industry*, pp. 20–21.

...,X_H); by-product production $(B_1,\ldots,B_J)$; materials recovery from residuals streams $(W_1,\ldots,W_K)$; modification and transport of residuals $(T_1,\ldots,T_L, V_1,\ldots,V_M)$; discharge of residuals $(D_1,\ldots,D_G)$; and sale of products $(Y_1,\ldots,Y_N)$. Corresponding to this division of the possible activities is the vertical division of the A-matrix and the objective function. Thus, for example, the objective function entries corresponding to the production activities are the costs of production; those corresponding to residuals discharges are zero unless some effluent charge is being levied.

The horizontal division of the A-matrix indicates broadly the type and form of the constraints included. Thus, each unit of production using process X_1 requires a vector of inputs represented by $(-\tilde{p}_{X_1})$. Total input requirements for production at level X_1 are $-\tilde{p}_{X_1} \cdot X_1$, and the input availability constraints simply say that

$$\sum_h (-\tilde{p}_{X_h}) \cdot X_h \geq -\tilde{P}$$

or that no input may be used beyond its level of availability as given by the applicable entries in the right-hand side or b-vector. The minus sign on inputs is used here primarily to emphasize the symmetry between traditional inputs and primary residuals generation. In practice, signs may be chosen for computational ease (for example, so as to avoid negative entries in the b-vector), so long as row consistency is maintained. The other constraints imposed include requirements that:

1. all products sold actually be produced
2. product output qualities conform to certain specifications
3. all residuals generated be accounted for by material recovery, by-product production, modification, transport, or discharge

The quality constraints, requirement 2, for production at levels $X_1, \ldots, X_H$ are shown in the simple form

$$\sum_h \tilde{q}_{X_h} \cdot X_h \gtrless Q$$

Because such constraints will generally be placed on the concentration of one or another substance in the product, and because final products often result from blending separate intermediate stocks, the actual construction of these rows will generally be more complicated. The method discussed below for including effluent concentration constraints may be applied to such product quality requirements.

Requirement 3, accounting for residuals generated, deserves more detailed discussion. The choices of by-product production, material recovery, modification, or transport option for handling some residual generated in production imply the generation of one or more of, what are called in the model, secondary residuals. These are basically of three types: that portion of a residual subject to modification which is not removed or altered by the modification process (such as the fly ash continuing up the stack after electrostatic precipitation); the new forms of residuals created by a modification process (such as the sludge from primary physical and secondary biological modification of waterborne oxygen-demanding compounds, or the water vapor and CO_2 from refuse burning); and the same residual at a *different place* (as when a liquid residual is piped elsewhere before discharge). The method of inclusion of secondary residuals and of the requirement that they, in turn, be modified, transported, or discharged is basically straightforward and might be called "row transfer." For example, consider an activity, say T_l, designed to remove some percentage b of residual i from the stream containing it, while in the process generating a quantity of residual, say r_{T1j}, of a secondary residual j all at a cost c_{T_l} per unit of input of residual i to the process (*not,* in this method, the cost per unit of i removed). The activity vector for this process would be

Row entry description	Modification activity (T_l) vector entries
Quantity of input of residual i to modification process	$+1$
Quantity of residual i *not* removed	$-(1 - b)$
Quantity of new residual j generated per unit of input of residual i	$-r_{Tlj}$
Objective function (cost of unit activity level)	$-c_{Tl}$

Thus, if this modification activity is operated at the level T_l sufficient to account for all the residual i assumed generated, the quantity $(1 - b)T_l$ of i would be unaffected, requiring no further modification or discharge, and the quantity $r_{T1j} \cdot T_l$ of the new residual j would now also have to be accounted for. But it will be instructive to follow this hypothetical residual i through the constraint matrix, using the notation in figure 4-A-1. In figure 4-A-2 we show the necessary matrix entries, the activity level designations, the right-hand side, and the objective function.

Modification activities l and m remove, respectively, the fractions b_l and b_m of the amount of residual i to which they are applied. Thus, for every unit of input of residual i to process l, $(1 - b_l)$ units remain for discharge, assuming—as we have here—no opportunity for further modification. Also, for every unit of input of residual i, the modification processes produce, respectively, r_{T1j} and r_{Tmj} units of a new residual, j. This residual, in turn, must be modified in process n, transported by process o, or discharged. And finally, both processes n and o produce further residuals requiring discharge. Writing out the indivdual row conditions for this example, we have:

$$-X_f(r_{Xfi}) - X_k(r_{Xki}) + T_l + T_m + D_i = 0$$
$$-T_l(1 - b_l) - T_m(1 - b_m) + D_{i'} = 0$$
$$-T_l(r_{Tlj}) - T_m(r_{Tmj}) + T_n + V_o + D_j = 0$$
$$-T_n(r_{Tnh}) + D_h = 0$$
$$-V_o(r_{Voo}) + D_g = 0$$

In the absence of effluent charges, the contribution to the objective function of this section of the problem is given by

$$-X_f(c_{Xf}) - X_k(c_{Xk}) - T_l(c_{Tl}) - T_m(c_{Tm}) - T_n(c_{Tn}) - V_o(c_{Vo}).$$

Activity Levels

	Residuals generation		Treatment			Transport	Discharge				
	$X_j \ldots X_k \ldots$		$T_l \ldots T_m \ldots T_n \ldots$			$V_o \ldots$	$D_i \ldots$	$D_{i'} \ldots$	$D_h \ldots$	D_g	
	X_j	X_k	T_l	T_m	T_n	V_o	D_i	$D_{i'}$	D_h	D_g	
row i	$-r_{X_{j_i}}$	$-r_{X_{ki}}$	$+1$	$+1$			$+1$				
.											
row i′			$-(1-b_l)$	$-(1-b_m)$				$+1$			
.											
row j			$-r_{T_{li}}$	$-r_{T_{mi}}$	$+1$	$+1$					
.											
.											
row h					$-r_{T_{nh}}$				$+1$		
.											
.											
row g						$-r_{V_{og}}$				$+1$	
Objective function	$-c_{X_j}$	$-c_{X_k}$	$-c_{T_l}$	$-c_{T_m}$	$-c_{T_n}$	$-c_{V_o}$	$*$	$*$	$*$	$*$	

The column of equalities on the right: $\left[\cdots = 0 \cdots \right]$

Figure 4-A-2. Residuals-handling in the linear model. Note: The asterisks indicate that prices (effluent charges) may be applied to the discharge activities. From: Russell, *Residuals Management in Industry*, p. 23.

One extension of this general method is worth specific mention for it allows us to deal with the common situation in which several streams (as, for example, process water streams from different processing units) contain a number of residuals in different proportions and are subject to several possible modification stages, each removing a particular proportion of each residual. The basis of this method is to use, in place of any single residual quantity, the quantity of the carrying stream as the variable in the required row continuity conditions. Only at the point of discharge are the concentrations of residuals used to obtain quantities of residuals explicitly. Thus, assume we are interested in two water streams of volumes V_1 and V_2, generated in production processes X_1 and X_2, respectively. Assume that both contain three residuals of interest, in concentrations O_{11}, O_{12}, O_{13} and O_{21}, O_{22}, O_{23}, respectively. Further assume that each stream may be discharged directly; subject to modification process A and then discharged; or subject to process A followed by process B and then discharged. Process A is assumed to remove fractions a_1, a_2, and a_3 of the three residuals; process B removes fractions b_1, b_2, and b_3 of the remaining quantities of the residuals. We neglect, for simplicity, all process inputs and any secondary residuals generated in A and B. Then the required matrix entries, and so forth, may be written as in figure 4-A-3. The total discharged quantities of residuals 1, 2, and 3 are simply obtained as the sum of discharges 1 through 6.

The most obvious problem with this approach involves economies of scale. If, in the program, any subset of the several streams may be chosen for modification and others discharged, we can never know in advance the total volume for which the facility must be designed. Hence, unit costs applied to A and B in the objective function must necessarily be arbitrary. We may guess at the probable volume, choose the largest possible volume, or adopt some other strategy, but whatever we do will have some effect, in turn, on the volume actually determined in the solution. As long as there are economies of scale, these guesses will tend to be self-fulfilling, because large assumed volume will imply low unit costs, which will in turn encourage the wide adoption of modification processes (and vice versa). In the specific model used, unit modification costs of petroleum refineries were based on the size of residuals modification plants that would be required to handle the entire volume of process water for the bench mark refinery. More subtle problems may arise because of nonconvexities introduced by variations in the modification per unit cost between A and B for one or more residuals.

Activity levels

Rows	A_1	A_2	B_1	B_2	D_1	D_2	D_3	D_4	D_5	D_6
Vol. 1 (from production)	+1	·	·	·	+1	·	·	·	·	·
Vol. 1 (from process A)	−1	·	+1	·	·	+1	·	·	·	·
Vol. 1 (from process B)	·	·	−1	·	·	·	+1	·	·	·
Vol. 2 (from production)	·	+1	·	·	·	·	·	+1	·	·
Vol. 2 (from process A)	·	−1	·	+1	·	·	·	·	+1	·
Vol. 2 (from process B)	·	·	·	−1	·	·	·	·	·	+1
Residual 1 discharged					$-O_{11}$	$-(1-a_1)O_{11}$	$-(1-b_1)(1-a_1)O_{11}$	$-O_{21}$	$-(1-a_1)O_{21}$	$-(1-b_1)(1-a_1)O_{21}$
Residual 2 discharged					$-O_{12}$	$-(1-a_2)O_{12}$	$-(1-b_2)(1-a_2)O_{12}$	$-O_{22}$	$-(1-a_2)O_{22}$	$-(1-b_2)(1-a_2)O_{22}$
Residual 3 discharged					$-O_{13}$	$-(1-a_3)O_{13}$	$-(1-b_3)(1-a_3)O_{13}$	$-O_{23}$	$-(1-a_3)O_{23}$	$-(1-b_3)(1-a_3)O_{23}$
Objective function	$-c_A$	$-c_A$	$-c_B$	$-c_B$						

Figure 4-A-3. Method of handling multiresidual streams subject to a variety of modification processes. Note: A_1 indicates the level of process A applied to stream 1. From: Russell, *Residuals Management in Industry*, p. 26.

Having explored in some detail how the linear response model reflects residuals-handling alternatives, we now describe how it can be used to investigate the response at an industrial plant to residuals management action by public authorities. We may use effluent charges, discharge quantity constraints, or even discharge concentration limits (or some combination) to determine the effects of these incentive measures on discharges, costs, production volume, and so forth. The effects of variations in *indirect* influences—such as process technology, requirements for output quality, available input quality, relative input prices, and so forth—can be determined, and under each set of these influences the impact of direct residuals management actions examined. Thus, if we have good reason to expect changes in any of the indirect influences, we are in a position to investigate the residuals generation and discharge pattern of the plant under the new conditions as well as under present conditions.

In figure 4-A-2, the place of effluent charges in the model is already indicated. Because discharges are explicit activities and not simply "slacks," effluent charges fit perfectly as the unit activity costs. If effluent charges, specific to residuals and to locations, were applied to the discharges in figure 4-A-2, the new objective function would include

$$- (D_i + D_{i'})\alpha_i - D_j\alpha_j - D_h\alpha_h - D_g\alpha_g$$

where α_i is the fee per unit of discharge of residual i, and so forth. If the charges were not specific to location, but only to type, we would have

$$-(D_i + D_{i'})\alpha_i - (D_j + D_g)\alpha_j - D_h\alpha_h$$

because, by hypothesis, residual g differed from residual j only in discharge location.

Constraints on discharge quantities are easily included by attaching additional rows. If we wish to constrain a specific discharge D_i, to be less than $\bar{D}_i$, we simply introduce a new row in which the activity (column) D_i has a $+1$ entry. Then, a constraint is put on the new row, requiring $D_i \times 1 \leq \bar{D}_i$. If several separate discharge activities all involve the same residual type, it is a simple matter to constrain their sum. If one wishes to constrain discharge concentrations (as milligrams of BOD_5 per liter of water discharged), the task is somewhat more complicated. Referring back to figure 4-A-3, consider the possibility of constraining the concentration of residual 1 to be less than $\bar{R}_1$. The total discharge of residual 1 is

$$Q = O_{11}D_1 + (1 - a_1)O_{11}D_2 + (1 - b_1)(1 - a_1)O_{11}D_3 + O_{21}D_4$$
$$+ (1 - a_1)O_{21}D_5 + (1 - b_1)(1 - a_1)O_{21}D_6,$$

while the total volume of the discharge stream is $D_1 + \ldots + D_6$. Thus, the concentration is

$$\frac{Q}{D_1 + \ldots + D_6}$$

and we wish to require that

$$\frac{Q}{D_1 + \ldots + D_6} \leq R_1$$

In this form, the constraint is unusable in the linear program, but if we clear fractions and subtract the right-hand from the left-hand side, we obtain

$$D_1(O_{11} - R_1) + D_2[(1 - a_1)O_{11} - R_1] + \ldots$$
$$+ D_6[(1 - b_1)(1 - a_1)O_{21} - R_1] \leq 0$$

Thus, the row entries become the differences between the actual and the desired concentrations, and the constraint simply says that the volume-weighted average of these differences must be less than or equal to zero. The difficulty with this technique is that exploration of alternative levels of the concentration constraint R_1 is cumbersome and time consuming, because, in general, the only way to proceed is to insert an entire new set of row entries in the vectors $D_1, \ldots, D_6$. (In the refinery, when dealing with a large number of alternative gasoline-blending stocks, or a large number of process-water streams, this becomes a significant problem.) Only if we are willing to require that the total volume remain the same for all solutions can we vary the level R_1 easily. To see this, let us define

$$Z \equiv D_1(O_{11} - R_1) + D_2[(1 - a_1)O_{11} - R_1] + \ldots$$
$$+ D_6[(1 - b_1)(1 - a_1)O_{21} - R_1]$$

so that the constraint requiring at most concentration R_1 becomes simply $Z \leq 0$. Further, we define

$$\bar{D} \equiv D_1 + \ldots + D_6$$

and require the total volume to be a constant. ($D_1, \ldots, D_6$ can vary individually.) Now, consider a new, tougher, concentration constraint $R_2 = R_1 - \Delta$. In general, the constraint for R_2 would be

$$D_1(O_{11} - R_2) + D_2[(1 - a_1)O_{11} - R_2] + \ldots$$
$$+ D_6[(1 - b_1)(1 - a_1)O_{21} - R_2] \leq 0$$

But this is equivalent to

$$D_1(O_{11} - R_1 + \Delta) + \ldots + D_6[(1 - b_1)(1 - a_1)O_{21} - R_1 + \Delta] \leq 0$$

or

$$D_1(O_{11} - R_1) + \ldots + D_6[(1 - b_1)(1 - a_1)O_{21} - R_1] + \bar{D}\Delta \leq 0$$

or finally,

$$Z \leq \bar{D}\Delta$$

Thus, if we guarantee constant total volume, a concentration constraint can be varied simply by varying the right-hand side.

A variety of indirect influences on residuals generation and discharge may be studied by manipulation of values of the objective function, the right-hand side, or the matrix of coefficients itself. Thus, in the objective function, any of the price or cost figures may, in principle, be altered and the effects observed, though in practice we may be interested only in the price of a key input (such as coal to a thermal-electric generating plant), a particularly important product (such as motor gasoline from a petroleum refinery), or of an actual or potential by-product (such as sulfur from the petroleum refinery). On the right-hand side, one may change input availabilities and output quantity requirements. Finally, advances in production or residuals-handling technology can be reflected by changing coefficients within the A-matrix itself. Such changes may take the form of introducing entire new columns to represent possible new processes. Another alternative is to change one or two coefficients within existing columns to reflect progress in a subprocess of a largely unchanged overall process.

Appendix 4-B
Flow Diagrams for Bleached Kraft Paper Production

This appendix consists of figures 4-B-1, 4-B-2, 4-B-3, and 4-B-4.

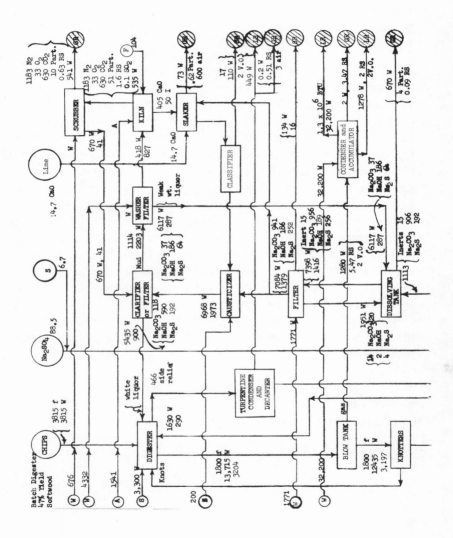

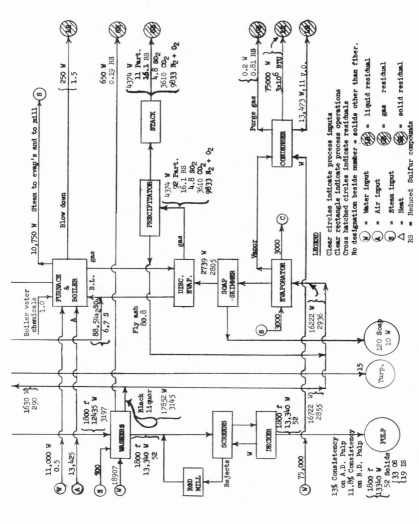

Figure 4-B-1. Kraft pulping #5

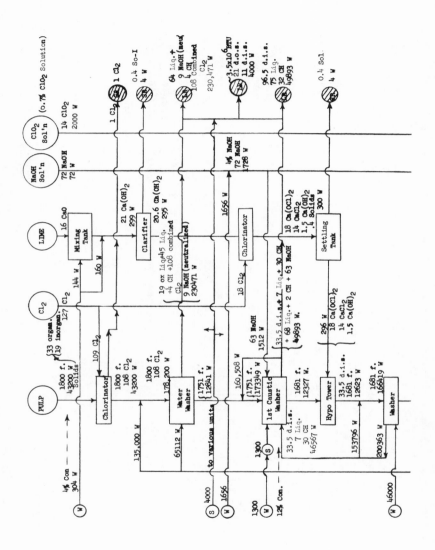

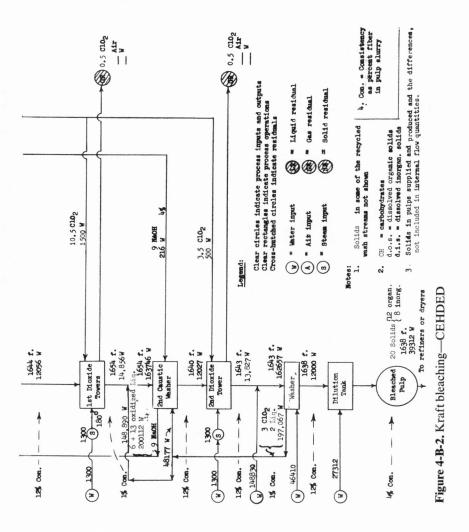

Figure 4-B-2. Kraft bleaching—CEHDED

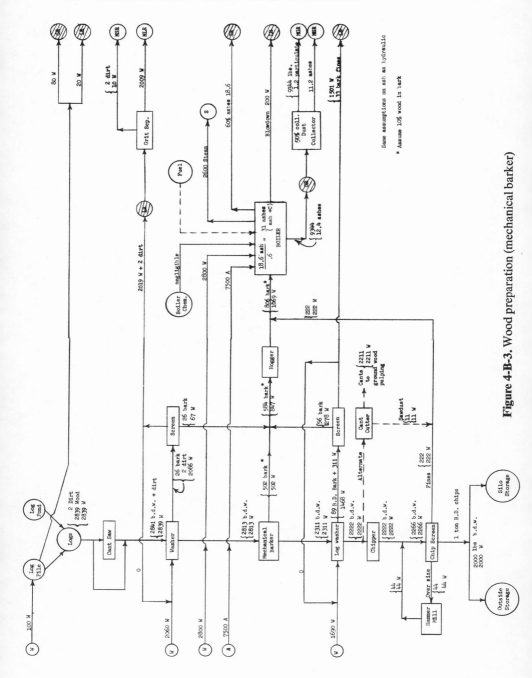

Figure 4-B-3. Wood preparation (mechanical barker)

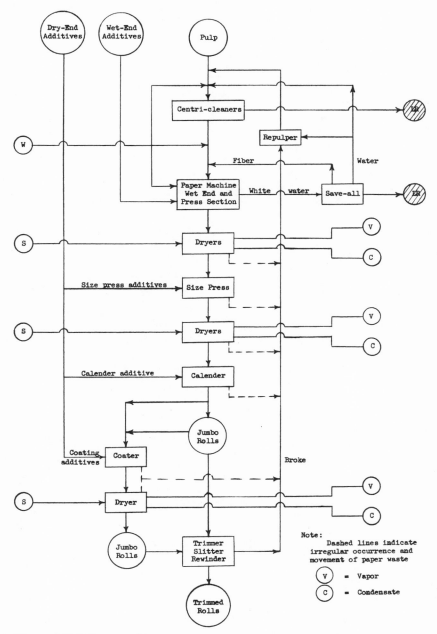

Figure 4-B-4. Paper making

PART THREE

Regional Scale Issues and Analyses

CHAPTER 5

Regional Residuals Management Models

It is useful to analyze integrated residuals management at the individual activity level because of the insights that such microlevel analysis can provide in and of itself, but also because it is essential to analyzing integrated residuals management on a regional level. Because actual decision making having to do with managing residuals discharged from multiple sources takes place in particular geographic areas, a regional REQM strategy involves a mix of incentives that affect both the individual dischargers (to induce them to reduce discharges) and the management agency itself (when it acts in a collective way to modify residuals and when it acts directly to modify environmental quality)—all this directed toward improving ambient environmental quality.

The analysis of integrated residuals management at the microlevel provides information on the possible responses of individual dischargers. The task of *regional* REQM analysis is to combine the alternative residuals management options for the individual activities with collective management options and the transformations produced by natural systems to develop an optimal (however defined) regional strategy, a strategy which considers explicitly the distribution of both the benefits and the costs of improving ambient environmental quality.

As discussed in chapter 2, when residuals are discharged, they are not only transported but they undergo various transformations in the natural systems which receive them. Accordingly, the impact that a given residuals discharge has on a particular receptor is a function not only of the amount and type of discharge but of where the discharger is located relative to the receptor and the specific characteristics of the receiving environmental media in that particular region. The latter include such factors as: air temperature, wind speed, the presence or absence of inversions; water temperature, water volume and flow, pre-existing biochemical composition; and land slope, infiltration capacity, soil type, and vegetative cover.

Thus, to analyze the economic, technological, institutional, and ecologic aspects of residuals management in a region (a term which we purposely leave vague for the time being) requires a quantitative model[1] that incorporates salient aspects of the economy, technological options for reducing residuals discharges, the behavior of the pertinent natural systems, and the location of receptors—and, for reasons explained previously, all this simultaneously for liquid, gaseous, solid, and energy residuals streams. One of the principal activities of the Quality of the Environment program at RFF over the past several years has been to construct one such model and make an illustrative application of it to a particular region.

The objective of this chapter is to describe the particular type of conceptual REQM model developed by the RFF team. This model is also known as the Russell-Spofford model. These terms will be used interchangeably in the following discussion. Applications of the model are described in chapters 6 and 7. Constructing a quantitative model of this type necessarily involves extensive use of mathematics. Since we are avoiding presentations of mathematical materials in the text, the reader interested in technical details of the regional application of the Russell-Spofford model will find a relatively complete, concise discussion of its mathematical structure in appendixes 5-A and 5-B.

The Model in General

A logical point of departure for understanding the kind of model needed for regional REQM is to consider first the programming model of residuals management in an industrial operation presented in the previous chapter. Programming models have long been used in some industries to make scheduling, product design, and other management decisions. A straightforward but important extension of this type of industry model is made by constraining not only the availability to the industrial plant of

---

[1] It should be emphasized that there is a wide range in the degree of sophistication of quantitative models, from fixed coefficient, "back of the envelope" type to complicated mathematical programming type models. The degree of model sophistication should be a function of the questions to be answered, the constraints in the given situation, and the analytical resources—personnel, time, data, computer facilities—available.

ordinary resources but also of environmental resources; that is, limitations are placed on the amount of residual materials and energy that may be discharged to the air, water bodies, and on the land.[2] By this means, the private internal costs imposed by policies such as effluent standards or taxes, which are meant to diminish discharges and their associated external costs, can be determined for alternative levels of application of these policies. This is useful general information for regional REQM when costs are a consideration—that is to say, nearly always.

But such models of activities, useful as they are in themselves, say nothing about the quality of the environment. To make this connection, models of the effect of residuals discharges on the pertinent natural systems are needed. Making this step is often desirable, because there exist, or can be developed, reasonably acceptable models for the atmosphere and water bodies (including groundwater aquifers), which translate a quantity of a residual discharged at some point x to an ambient concentration at some point y. Similar models are possible for residuals-carrying runoff from land surfaces.

In constructing natural systems models, environmental media (by which we mean the atmosphere and water bodies) are usually divided conceptually into a number of grids or blocks within which environmental conditions are taken to be uniform. Each block is represented by differential equations describing the rates of flow of materials, energy, or both, across the boundaries of the blocks as a function of conditions within the blocks and the rates of biochemical and physical transformations or decay of these materials within the blocks as functions of the same conditions.

By means of such models, it is possible to include ambient environmental quality constraints in addition to effluent constraints in the regional analysis (or environmental quality constraints can be considered alone). Moreover, the dispersion-type natural system models are not limited to a single source of discharge. They can translate discharges from multiple point, area, and line sources to concentrations at multiple receptor points.[3] Natural systems models are discussed in more detail in appendix 5-B.

As part of the RFF research program on REQM, a new aquatic ecosystem model of an estuary was developed by Robert Kelly. This model differs from the traditional dispersion-type model in that, in addition to

[2] Similar extensions can of course be made to models of activities other than industrial, that is, commercial, residential, agricultural, silvicultural, and mining.

[3] This paragraph is not meant to imply that either it is always necessary to develop such models or that it is always feasible to do so.

converting weights of materials and quantities of thermal energy at points of discharge into concentrations of those materials at specified locations, it also converts them into measures of effects on living populations, such as zooplankton, algae, and fish. The last two are of more direct interest to environmental quality policy making than simple concentrations of residuals in the environment. Although it involves a large amount of computation, the availability of natural systems models therefore makes it feasible to minimize, for example, the overall private cost of achieving a specified ambient standard, or set of standards, either in terms of concentrations of materials or biomass of living species, or both, in a region. This is accomplished by means of a large regional mathematical model which encompasses aspects of the behavior of the actual environmental systems (water bodies and the air mantle). Thus there exists, at least in rudimentary form, an economic-ecological model which in principle should be useful for REQM in, say, a river basin, a metropolitan airshed, or a groundwater basin.

But these seemingly straightforward extensions of well-established management science models raise fundamental questions and pose basic difficulties in trying to provide systems analyses useful for regional REQM. There would appear to be four main questions: (1) How far should the explicit incorporation of the various sorts of interdependencies among residuals and among media be carried? (2) How should the outputs (external costs avoided) of the management systems be evaluated, or should this even be attempted? (3) What is the appropriate objective or objectives? (4) How can and how should the model interact with actual and potential political structures?[4] The first question is particularly difficult in connection with REQM problems because of the genuine complexities involved and the prominence of "ecological thinking" in the field. Hence, before discussing the particular regional REQM model used, some more general comments on this problem are relevant.

Simulation Versus Formal Mathematical Optimization

The "frog in the hole in the bottom of the sea" chain of reasoning of some ecologists has led to visions of the REQM problem which push it beyond the bounds of any kind of successful modeling. Some years ago in an

[4] There is of course the important subsidiary question of how detailed the various submodels (activities and natural systems) should be.

article on operations research, Robert Dorfman put the general point very well.

> As a result of complexity the operations analyst, like every other worker, lives nearly always near the end of his tether. He simplifies his problems as much as he dares (sometimes more than he should dare), applies the most powerful analytic tools at his command and, with luck, just squeaks through. But what if all established methods fail, either because the problem cannot be forced into one of the standard types or because, after all acceptable simplifications, it is still so large or complicated that the equations describing it cannot be solved? When he finds himself in this fix, the operations analyst falls back on "simulation" or "gaming."[5]

All models are simulations in that some exogenous conditions are chosen, the model is run with the assumption that it mimics some part or parts of the real world and the behavior of the actors therein, and an "answer" is spewed out. In simulation models, relationships are given in the form of linked equations that are not solved analytically but computed numerically. Formal mathematical optimization models, however, are constructed to require analytical solutions, and, therefore, if there are a number of simultaneous relationships involved, they usually have to be primarily of a linear variety in order to make their solution feasible. To obtain this linearity, it is necessary to work with more or less good approximations of the actual functional relationships involved. Computer simulation can handle complex relationships beyond those manageable in mathematical optimization models, such as linear programming models. But one result of such ability is that much of any systematic method for finding a better answer in relation to the objective is lost, and the problem of choosing among alternative outcomes of the simulation can easily become very difficult.

Consider a small simulation model in which there are twenty-eight variables (an actual regional REQM model may easily have many hundreds, as chapter 7 indicates), each of which may be set at any one of three levels. There are then 3^{28} possible systems, assuming all variables are independent. This amounts to approximately 23,000 billion. If it takes one minute of computer time to simulate each design, it could require about 50 million years to complete the simulation. Of course, no simulator would attempt a complete enumeration of outcomes in a large problem, but this calculation does suggest the complexities to be faced. The same complexity faces the modeler using mathematical optimization, but math-

[5] Robert Dorfman, "Operations Research," *American Economic Review* vol. 50, no. 4 (1960) p. 575.

ematical optimization provides a systematic method for finding a "better and better" answer in relation to the chosen objectives. (However, complications arise when there are multiple optima, as is often the case for complex, regional REQM systems.)

Simulation may be appropriate if informed judgment can readily yield a relatively few alternative systems to analyze, or if there are sufficient constraints so that only a few possible options are to be analyzed. Neither of these conditions is likely to occur when large REQM systems are involved. Alternatively, simulation can be supplied with objective functions and one or another form of sampling used to generate a "response surface." But the principles of sampling for this type of problem are not well understood, and providing an adequate sample may itself be a very difficult problem if the number of variables and alternative levels is great and the response surface is irregular.

Be that as it may: (1) even with mathematical optimization there is never a guarantee that the "best" answer will be found if there are multiple optima; and (2) the system may be sufficiently complex so that optimization may not be very efficient and—in the limit—may amount to nothing more than "brute force." Thus the line between mathematical optimization and simulation may virtually disappear. But clearly there are trade-offs between the two in terms of the types of output information produced.

The question of simulation versus mathematical optimization can be summarized by indicating the considerations to be taken into account in the choice of analytical approach, keeping in mind that combinations of simulation and mathematical optimization are possble.

1. Mathematical optimization imposes a valuable discipline on the modeler and the modeling process.
2. If ambient environmental standards are the targets to be achieved by the REQM strategy and there are a large number of dischargers, a large number of receptors, and a large number of possible discharge reduction options, then mathematical optimization will usually be the reasonable way to proceed.
3. If there are only a few major dischargers and a few options for reducing residuals discharges, simulation may be easier and efficient enough.
4. If there are many similar sources of residuals discharges, but only one or two options for reducing discharges at each source, and the objective is defined in terms of required reductions or percentage of

reduction at the sources (rather than in terms of ambient environmental quality), simulation again may be easier and efficient enough.

5. Large mathematical optimization models are expensive and difficult, unless linear in nature. However, the assumptions necessary to make a linear model may do great violence to the real world, for example, result in unacceptable inaccuracies.

6. Large, nonlinear mathematical optimization models almost always (an estimated 90 percent of the time) have multiple optima.

7. Synergistic and antagonistic effects are more difficult to handle in mathematical optimization than in simulation models.

8. Economies of scale or analogous increasing-return situations cannot be handled satisfactorily by any mathematical optimization technique.

Finally, it should be remembered that analyzing any complex REQM system by any approach involves obtaining, arranging, and handling large amounts of empirical data. In many cases the major problem is that of obtaining empirical data on activities—production processes, residuals generation, and residuals discharge reduction options and their costs—and on the effects of discharges on ambient environmental quality.

On the basis of such considerations, the modeling team at RFF concluded that, with all the limitations, some form of mathematical optimization model was more likely than a simulation one to be useful for decision making in the particular REQM case study described in chapter 7. Mathematical optimization models can also, of course, incorporate simulation submodels to provide descriptive linkages in a set of nested models, and the one developed by the team does (the aquatic ecosystem model mentioned above is a simulation model). But mathematical optimization does not sacrifice specifying a criterion function and thus results in an orderly approach to the optimal solution, for whatever objective or objectives are included in the criterion function. Selected parameters of the mathematical optimization model can be varied and new solutions found, and this would almost always be desirable in real decision-making situations, but the hallmarks of mathematical optimization models—an explicit objective function with constraints and a specific solution procedure—are not thereby abandoned.

Clearly it is necessary to recognize that regional REQM models can never be comprehensive in the sense that they consider all linkages and all alternatives. Therefore, great care must be taken to specify what aspects of reality are or are not included. In this connection a well-functioning market system can be of great help—a fact often ignored in the more ecologically oriented models—in that many aspects of resource use can be excluded from explicit consideration in REQM models on the grounds that they are appropriately handled by the market exchange system. The interface between these processes and the model as such is through the system of value weights generated by the market, that is, prices.[6] The model itself can then focus explicitly on those aspects of resource use where market exchange is known to fail seriously as an allocative device, for example, with respect to allocation of the common property resources of air, water, and associated ecological systems.[7]

Stochastic and dynamic elements are likely to be particularly difficult to handle adequately in mathematical optimization models. Inability to do so represents an important deficiency. Sensitivity analysis can help some, but the wise modeler will never let himself be lured into thinking that either on conceptual or empirical grounds will his models provide a complete basis for decision making in a field as complex as REQM. The mathematical models must be viewed as potentially helpful tools that constitute one element, albeit a major one, in a decision-making process —tools which can reveal nonobvious impacts of "common sense" policies and go some distance in quantifying them. They must be built and used because the inherent logic of the problem demands it, and we have nothing better.

The REQM modeling research at RFF began with a conceptual model developed by Clifford Russell and Walter Spofford. The original framework and a didactic application to a small hypothetical region have been described in detail.[8] We present here the basic features of that framework.

---

[6] This is not to say that these links—prices—can always be taken to be exogenous constants. They may be exogenous variables whose value depends upon the state of the REQM system. In this case a management model might be linked with an input–output model to capture secondary stresses in the system.

[7] As the discussion of collective decision making in chapter 6 makes clear, there are multiple objectives in most, if not all, decision contexts. It is highly unlikely that all of these can be incorporated in a single objective function. Consequently, the results of mathematical optimization models represent only part of the information that goes into the decision-making process.

[8] Clifford S. Russell and Walter O. Spofford, Jr., "A Quantitative Framework for Residuals Management Decisions," in Allen V. Kneese and Blair T. Bower, eds., *Environmental Quality Analysis: Theory and Method in the Social Sciences* (Baltimore, Johns Hopkins University Press for Resources for the Future, 1972).

Because the rationale for many of its features has been developed in previous chapters, and also exists in the appendixes to this chapter, we can be concise here. How this conceptual model was actually applied to a real region is described in chapter 7.

The Conceptual Russell-Spofford Model

The Russell-Spofford REQM model has the following components:

residuals generation and discharge models (activity models) which describe the factors influencing the generation, modification, and final discharge to the environment of liquid, gaseous, solid, and energy residuals from individual production and consumption activities, and the relevant production and residuals modification costs for one or more sets of prices of factor inputs to the activities

models of collective residuals modification activities, which detail the inputs, outputs, and costs of activities (such as municipal incinerators, municipal sewage treatment plants, used-paper processing plants) that modify residuals collected and transported from several residuals-generating activities

environmental modification models, which describe the options available for directly improving the assimilative capacity of the environment, such as low-flow augmentation and instream aeration, and their costs

natural systems models, which translate the time and spatial pattern of residuals discharges into the resulting time and spatial pattern of ambient environmental quality in terms of residuals concentrations and population sizes or densities of biological species

damage functions, which relate the time and spatial patterns of ambient residuals concentrations to the resulting impacts on receptors—man, animals, plants, structures—in physical, biological, physiological, and economic terms

evaluation and objective functions, which specify how the inputs and outputs are to be evaluated and the structure and components of the objective function.

The second and third components are self-explanatory. Further comments on the other four components follow.

Activity Models

Like the regional REQM model as a whole, activity models deal simultaneously with liquid, gaseous, solid, and energy residuals, recognizing explicitly the physical and economic links among them. They "recognize," for example, that removing waterborne organic residuals by standard sewage treatment processes creates a sludge which, in turn, represents a solid residuals problem; the sludge must either be disposed of on the land or burned, the latter alternative generating airborne particulates, other gaseous residuals, and a remaining solid residual. These models relate inputs and outputs of production processes and consumption activities at specified locations within a region, including: the unit amounts and types of residuals generated in the production of each product and in residential and commercial activities; the costs of changing these residuals from one form to another (as gaseous to liquid in the scrubbing of stack gases); the costs of transporting the residuals from one place to another; and the costs of any final-discharge-related activity such as landfill operations. Choices are possible among production processes, raw material input mixes, by-product production, materials recovery, and in some cases product mix and product specifications—all of which can reduce the total quantity of residuals requiring handling and disposal. That is, the residuals generated are assumed to be not fixed, either in form or in quantity. These models also allow for choices among modification processes and hence among the possible forms of the residual to be disposed of in the natural environment and among the locations at which such discharges are accomplished.

Natural Systems Models

Natural systems models—physical, chemical, and ecological—describe the fate of various residuals after their discharge into the natural environment. They may be thought of as transformation functions operating on the vector of residuals discharges and yielding another vector of ambient concentrations at specific locations in the area and, in the instance of the aquatic ecosystem model, as having an effect on living species as well as on ambient concentrations in the water body. Between discharge point and receptor locations, the residual may be diluted in the relatively large volume of air or water in the natural world, transformed from one form to another (as in the decay of oxygen-demanding organics), accumulated or stored (as in the accumulation of organics in benthal deposits), and, of course, transported to another place. Fortunately, for many situations, the equations characterizing the transformation of residuals between discharge and receptor locations reduce to simple linear forms for steady state deterministic conditions, so that the linkage can be made via a coefficient.[9] The models which explicitly incorporate effects on living things are more complex and, as already mentioned, are in the form of simulation programs.

Damage Functions

The Russell-Spofford conceptual model includes a set of receptor-damage functions relating the concentrations of residuals in the environment and the effects on living things to the resulting damages, whether these are sustained directly by humans or indirectly through effects on material objects or on such receptors as plants or animals in which human beings have a commercial, scientific, or aesthetic interest. For the conceptual model, these functions relating concentrations and effects on species to damages are assumed to be expressed in monetary terms. The derivatives of these functions represent marginal damages, which could be summed for affected receptors for any particular discharge and thus provide an estimate of the appropriate effluent charge for that level of discharge by the specific discharger (given no synergistic and antagonistic relationships among residuals).

Assuming the damage functions are expressed in monetary terms, we can then explain the linkage between the components of the model and the method of optimum seeking in the following manner. The regional REQM model can be constructed as a linear programming model, which is solved initially with no restrictions or prices on the discharge of residuals. Using the resulting initial set of discharges as inputs to the natural systems models and the resulting ambient concentrations and effects on living species as arguments of the damage functions, as indicated above, the increment in damages can be determined as the change in damages associated with a unit change in a specific discharge. These incremental damages may then be applied as interim effluent charges on the discharge activities in the linear model, and that model solved again for a new set of production, consumption, modification, and discharge activities. The

[9] It should be noted, however, that physical, chemical, and biological interactions among residuals in the environment cannot be handled in this simple manner.

procedure is repeated until a position close to the optimum is found, that is, convergence occurs.[10]

In the real world, not all damage functions can be expressed in monetary terms, but let us assume that such damage functions could be found for all residuals, so that in fact a full-scale optimization is possible in the sense that no arbitrary ambient or discharge standards have to be specified. This leads us to another of the basic questions mentioned earlier. *Should* decision makers, that is, representatives of the public—if there are any such representatives, with appropriate responsibilities, a matter to which we will return subsequently—accept this "optimum" result as *the* basis for policy? It is pertinent to discuss this question because the conceptual model we are describing is in the format of traditional economic benefit-cost analysis which, in principle, aims at maximizing the efficiency or net benefit of the whole system. A necessary condition for this is that incremental damages be equated with incremental costs of management, as the solution just described would do. The two closely related issues raised by this general objective are: measurement of the value of inputs and outputs and the exact specifications of the objective function—the second and third questions raised earlier—to which we now turn.

Evaluation and Objective Functions

As we saw in chapter 1, modern welfare economics (that branch of economic theory which examines the normative properties of market economic systems) concludes that: (1) if preference orderings of consumers and production functions of producers (the functions defining physically efficient combinations of inputs and outputs) are independent and their shapes appropriately constrained; (2) if consumers maximize utility subject to given income and price parameters; and (3) if producers maximize profits subject to those price parameters—a set of prices exists such that no individual can be made better off without making some other

[10] The regional REQM problem could also be stated in a completely general manner not suitable to numerical solution but perhaps easier to understand. The objective is to maximize, for a region, a complicated economic criterion function reflecting the costs of regional production, the benefits from regional consumption, the costs of residuals modification, and the external damages resulting from residuals discharges, with allowance for the "assimilative" capabilities of the regional environment. In this form, the regional residuals management problem is a general nonlinear programming problem, with both objective function and some constraints being nonlinear. For more detailed explanations of the general structure of the Russell-Spofford model, see Russell and Spofford, "A Qualitative Framework."

individual worse off. For any given distribution of income, this is an efficient state. Given certain further assumptions concerning the structure of markets, this Pareto optimum can be achieved via a market pricing mechanism and voluntary decentralized exchange.

The fundamental conceptual basis of benefit–cost analysis is this theorem (although it is not illogical to defend such analysis on more intuitive grounds). It is also the normative basis for the Russell-Spofford model under discussion. What benefit–cost analysis does, essentially, is to attempt to simulate a market result in which, for one reason or another, voluntary exchange does not exist and the resources allocation decision is made through the agency of government. For the most part the inputs to such a management program—a project or a system of activities—are private goods purchased by the activities implementing the program, and it is common practice to use market costs as acceptable estimates of social costs. Even this requires a long string of assumptions, for example, that all markets are in competitive equilibrium and there is no unemployment of resources in the economic system. Methods for relaxing some of these assumptions have been developed, but for present purposes we assume that costs are acceptably measured so that the discussion can be turned to the far more serious problem of benefit measurement.

In conventional benefit–cost practice, benefits are measured by making estimates of what consumers would be willing to pay in a market for alternative levels of the publicly provided good or service. As we mentioned, this means simulating a market result where none exists. For example, a mathematical function could be developed which relates incremental improvements in air quality to the "willingness to pay" for them. Unfortunately, and especially when applied to REQM problems, this approach presents extremely difficult practical problems of measurement and requires some very heroic assumptions about distributional impacts.

In the case of private goods bought and sold in competitive markets, the matter of income distribution can be separated rather nicely from the problem of resources allocation. But the outputs of REQM actions are, as we have remarked several times, always public goods. This means that they are provided to large groups of people simultaneously (although different groups within a region, such as a metropolitan area, may receive significantly different amounts). For example, a reduction in sulfur dioxide (SO_2) may affect major parts of an entire city, and the amount I get if I am a resident of that part of the city is about the same as you get if you are a resident of the same part, and my breathing of the cleaner air does not deprive you of doing it simultaneously. This is different from a private good, in which case, if I eat it, you cannot. As stated in chapter 1, the technical name for this phenomenon is "jointness in supply."

Thus the income distribution problem associated with these public goods is particularly difficult because the use of the good cannot be differentiated among users on the basis of their voluntary choices in markets. When the supply of a public good (for example, air quality in a city) changes, both efficiency (resources allocation) and distribution (of goods and services) are inevitably affected, as we have seen. Consequently there is, in general, no way to be sure that equating incremental cost with the sum of incremental willingness to pay will yield a welfare maximum (this, we repeat, is the necessary condition for maximum net benefit). The preexisting distribution of income cannot be maintained except through an elaborate system of side payments tailored to each individual. Such a system of payments is a practical impossibility. This is unlike an economy consisting solely of private goods where equality of marginal cost and price (marginal willingness to pay) can be unambiguously shown to be a necessary condition for a welfare maximum, in the sense of the central theorem. In such a private-goods-only economy, allocative and distributional questions can be neatly separated.

The distributional effect associated with the provision of public goods has been a hard problem for applied public economics, and several procedures have been used to circumvent it. One way is to assume that the marginal utility of income (the satisfaction a person gets from his last dollar of income) is the same for all individuals. This effectively wipes out distributional considerations, because total utility (or welfare) is unaffected by how the output is distributed among recipients. But most economists regard this assumption as grossly unrealistic.

The assumption most often made implicitly or explicitly by applied economists is that it is a mistake to consider individual public goods situations in isolation. Rather, the whole complex of public goods should be considered. Some will affect one group adversely and another group favorably and, on another occasion, vice versa. Thus, there will be many cancellations of distributional effects, and it then follows that the society that makes its decisions based on the efficiency criterion (maximum net benefit disregarding distribution) will be one in which most people will finally be better off than one in which criteria are used which foreclose efficient solutions. It is further assumed that public goods are a rather small part of the economy and that private goods are allocated (through tax and subsidy policy) in an ethically sanctioned way. No economist that we know of is completely happy with this set of assumptions, but it does provide a somewhat defensible basis for testing the efficiency of public decisions.[11]

The scope and difficulty of the empirical measurement required, plus our inability to handle distributional questions in an entirely defensible way, make it impossible to achieve Pareto-optimal allocations of public goods by simulating markets. Therefore, political entrepreneurs who are concerned with general taxation (the allocation between the public and private sectors) and the allocation of government budgets must, in a democratic society, assume general responsibility for deciding about the provision of public goods. Assessing willingness to pay in selected cases may, however, play a very useful role in helping to focus the political process, particularly when it can be shown that large overall efficiency gains are potentially possible. Such evidence gives incentives to political entrepreneurs and ammunition to the public. Moreover, and most unfortunately, the economic technician and the planner or management scientist are often forced to make allocative decisions on the basis of partial evaluations because no proper political institution representing the people involved exists to make such decisions.[12] An example of such an allocative decision is the question of who receives how much of what kinds of environmental quality and who pays how much. In recent times, one result of this has been strong public resistance, through nongovernmental means, when an effort is made to implement an allocation decision. This takes the form of milder or more aggressive forms of "sabotage," often through the medium of the courts.

In the final analysis, the evaluation and allocation of public goods necessarily is a political problem in any society. This is true however useful the necessarily partial benefit–cost analyses made by technicians may be in aiding and even guiding the political process (or collective choice process as we like to call it). Final dependence cannot be put on disinterested evaluations of efficiency, even if there were a practical, poli-

---

[11] Note that the same problem exists in nonmarket economics, where publicly provided goods (not necessarily "public goods" in the strict sense) comprise a larger portion of the economy and where prices are often structured to achieve specified social goals, rather than economic efficiency.

[12] It is also true, however, that, even where reasonably representative governmental agencies exist, the politicians often fail to provide adequate guidance to the professional staffs.

tically sanctioned way of doing it, which there is not. To repeat, this is because such evaluation is necessarily partial and because it does not resolve adequately the distributional problems inherent in public goods situations.

The general point that evaluation of public goods is in the final analysis a political problem is clearly valid. But one cannot reasonably argue, we think, that our present political structures are adequate for solving environmental systems problems. This is surely one reason that the *value* problems associated with REQM have so often been converted into bogus technical problems. But to see what governmental arrangements might be more suitable, we must develop at least the rudiments of a theory of collective choice. As indicated in chapter 2, one element of the RFF Quality of the Environment program in recent years has been the development of such a theory and its application to integrated REQM in a region. In the next chapter we discuss some aspects of collective choice theory and illustrate how the political model was applied in the didactic example for a hypothetical region.

Appendix 5-A
Management Model Formulations and Optimization Scheme

In this appendix we describe the mathematical structure of the overall Russell-Spofford model used in the application described in chapter 7. In appendix 5-B we discuss in some detail natural systems models, and in particular, the one incorporated in the application.

Model Formulation

The objective function of the model is in the form of a cost minimization function. This does not mean that the management decision would be related to a single computed optimum. The model is formulated in this way for computational convenience. The coefficients in the objective function include: all the opportunity costs of traditional production inputs; all liquid, gaseous, and energy residuals modification costs; all collection, transport, and landfill costs associated with the handling and disposal of solid residuals; and all costs of directly modifying ambient environmental quality.

AUTHORS' NOTE: This appendix is based on material prepared by Walter O. Spofford, Jr., and Clifford S. Russell.

There are, basically, four types of constraints in the management model: (1) traditional resource availability (inequality); (2) continuity, or mass balance, relationships (equality); (3) constraints (inequality) on the increased costs of certain consumer services (for example, electricity, sewerage, solid residuals disposal); and (4) ambient environmental quality (inequality) constraints. The last, which involve the use of environmental models, are employed to constrain the levels of ambient environmental quality. They present the most difficulty and are therefore discussed in the greatest detail.

The regional REQM optimization problem is stated formally as

$$\min\{F = C(X,R)\} \tag{1}$$

$$\text{subject to } g_i(X) = 0 \qquad i = 1, \ldots, m < n - q \tag{2}$$

$$g_i(X) \geq 0 \qquad i = m + 1, \ldots, p \tag{3}$$

$$h_i(X) = R_i \qquad i = 1, \ldots, q \tag{4}$$

$$R_i \leq S_i \qquad i = 1, \ldots, q \tag{5}$$

$$R_i \geq 0 \qquad i = 1, \ldots, q \tag{6}$$

$$X_i \geq 0 \qquad i = 1, \ldots, n \tag{7}$$

where $C(X,R)$ is, in general, a nonlinear objective function; $g_i(X) = 0$, $i = 1, \ldots, m$, is a set of linear equality constraints; $g_i(X) \geq 0, i = m+1$, $\ldots, p$, is a set of linear inequality constraints; $h_i(X) = R_i, i = 1, \ldots, q$, represents a set of environmental functions which relate residuals discharges to ambient concentrations of residuals (discussed in detail in appendix 5-B); $X_i, i = 1, \ldots, n$, is a vector of decision, or management, variables, including production levels, residuals modification processes, and residuals discharges; $R_i, i = 1, \ldots, q$, is a vector of ambient levels of residuals concentrations and population sizes of species; and $S_i, i = 1, \ldots, q$, is a vector of ambient environmental quality standards (for example, concentrations of sulfur dioxide and particulates in the atmosphere, and of algae, fish, and dissolved oxygen in the water).

Some of the necessary environmental functions, $h_i(X) = R_i$, are available in linear form, for example, the air dispersion relationships and Streeter-Phelps-type dissolved oxygen models. Others are available only in nonlinear analytical form, while still others are available in various other forms. For nonlinear aquatic ecosystem models, no analytical expressions—either linear or nonlinear—of the form $h(X) = R$ are available. The variables R_i, $i = 1, \ldots, q$, are expressible only as a set of implicit nonlinear functions and, hence, simulation and other iterative techniques must be used to compute their values. (These various environmental model types are discussed in appendix 5-B.) In general, the environmental constraint set, equation (4), represents a variety of functional forms, many of which are difficult, or even impossible, to deal with using traditional mathematical programming techniques.

Because the optimization scheme (described below) requires that all the constraints be linear, the environmental relationships are removed from the constraint set and dealt with in the objective function. This modification of the problem requires the use of the penalty function concept discussed below.

The new optimization problem may be stated formally as

$$\min\{F = C(X) + P(X)\} \tag{8}$$

$$\text{subject to } g_i(X) = 0 \qquad i = 1, \ldots, m \tag{2}$$

$$g_i(X) \geq 0 \qquad i = m + 1, \ldots, p \tag{3}$$

$$X_i \geq 0 \qquad i = 1, \ldots, n \tag{7}$$

where

$$P(X) = \sum_{i=1}^{q} p_i[S_i, R_i = h_i(X)] \tag{9}$$

and where $p_i(S_i, R_i)$, $i = 1, \ldots, q$, are the penalty functions associated with exceeding the environmental standards, S_i, $i = 1, \ldots, q$.

Although the optimization scheme used requires only that those constraints (environmental relationships) that are not of the linear form $R = A \cdot X$ be removed, we note from the formulation of the new problem, equations (8), (2), (3), and (9), that even the linear environmental models have apparently been removed as constraints. This is optional and depends upon the model formulation and its size. If model size (the number of rows and columns) is of no consequence and if the entire management model is contained within a single linear program, it is more efficient to keep the linear environmental relationships as part of the constraint set. If size is a problem, or if it is otherwise desirable to divide the management model up into a number of smaller linear programs to be solved sequentially, disposition of the linear environmental models is not as straightforward.

In order to separate a large linear programming problem into a series of smaller ones, either the smaller ones must be completely separable in the sense that none of the original constraints relates variables which are now located in different linear programs, or one of the available decomposition techniques must be employed. No matter how the larger linear program is subdivided, the environmental relationships, which involve all the liquid and gaseous residuals discharges throughout the region, invariably link the smaller programs. Removing these relationships from the constraint set and placing them in the objective function is similar to the decomposition method known in the literature as "price coordination" (Lagrange method) or "dual feasible decomposition."

The scheme used to eliminate environmental relationships from the constraint set and still meet the environmental quality standards, S_i, $i = 1,\ldots,q$, is known as a penalty or exterior point method (as opposed to a barrier or interior point method). The term derives from the fact that throughout the optimization procedure the vector of standards S, equation (5), is allowed to be violated, but only at some "penalty" to the value of the objective function. The objective of the approach is to make this penalty severe enough in such a way that at the optimum the standards will be satisfied, within some tolerance, and the penalties reduced to essentially zero.

Because in the optimization scheme used the gradient is evaluated, it is necessary that the objective function equation (8) be continuous and have continuous first derivatives. A quadratic penalty function of the following form satisfies these requirements

$$p_i(X) = \max \, [(h_i(X) - S_i), \, 0]^2 \qquad i = 1, \ldots, q. \tag{10}$$

For computer applications, equation (1) may be written more conveniently as,

$$p_i(X) = \left[\frac{(h(X) - S) + |(h(X) - S)|}{2} \right]^2 \qquad i = 1, \ldots, q, \tag{11}$$

a form which gives $p(X)_i = 0$ when $h_i(X) < S_i$

The major difficulty with the penalty function expressed above as equation (10) is that, in general, it is not steep enough in the vicinity of the boundary and, consequently, the "unconstrained" optimum is apt to lie substantially outside the feasible region. However, a slight modification to the P function remedies this situation. If $r > 0$, and $p_i(X) \neq 0$, the new penalty function

$$\frac{1}{r_i} p_i(X) \tag{12}$$

approaches infinity as $r_i \to 0$. Specifying a sequence of decreasing values for r has the effect of moving the unconstrained optimum closer to the boundary of the feasible region. From a computational point of view, it is sufficient that r only be made small enough to ensure that the unconstrained optimum is within a preselected distance of the boundary.

This procedure may be explained heuristically as follows. Given $r^{(1)} > 0$, where $r^{(1)}$ is usually chosen to be 1, $X^{(1)}$ is solved for using one of the available maximization techniques. Then, using $r^{(2)}$, $X^{(2)}$ is solved for, again using one of the available maximization techniques, and so on. When $X^{(k)}$ satisfies (within a specified tolerance) the original constraint set, the procedure is terminated, as this is also the optimal point for the original problem.

The Optimization Procedure

Thus far, a formal mathematical statement of the regional REQM problem has been presented, a penalty function method for eliminating all but the linear constraints discussed, and the problem restated in terms of a nonlinear minimum cost objective function and a linear constraint set, but nothing has been said about the nonlinear optimization procedure used. We now turn to this.

The procedure used is analogous to the gradient method of nonlinear programming. The technique consists of linearizing the response surface in the vicinity of a feasible point $X^{(k)}$. To do this, a tangent plane is constructed at this point by employing the first two terms of a Taylor's series expansion (up to first partial derivatives). This linear approximation to the nonlinear response surface will, in general, be most accurate in the vicinity of the point $X^{(k)}$ and less accurate as one moves farther away from this point. Because of this, a set of "artificial" bounds is imposed on the system to restrict the selection of the next position along the response surface to that portion of the surface most closely approximated by the newly created linear surface. The selection of the appropriate set of artificial bounds is analogous to choosing a step size in other gradient methods of nonlinear programming.

Because the newly created subproblem is in a linear form, it is possible to make use of standard linear programming techniques for finding a new optimal point, $X^{(k+1)}$. This point locates the minimum value of the linearized objective function within the artificially confined area of the response surface. Because, in general, the linearized surface will not match the original nonlinear surface, the original nonlinear objective function must be evaluated at this point to determine whether or not this new point $X^{(k+1)}$ is, in fact, a better position than the previously determined one $X^{(k)}$. That is, the following condition must be satisfied

$$F[X^{(k+1)}] > F[X^{(k)}]$$

If this condition is satisfied, a new tangent plane is constructed at the point $X^{(k+1)}$, and a new set of artificial bounds is placed around this point. As before, linear programming is employed to find a new position $X^{(k+2)}$, which minimizes the linearized objective function, and so on until a local optimum is reached.

As has been pointed out, this optimization procedure requires that the objective function be linearized at a point $X^{(k)}$. This is done according to the formulation explained in the following section.

Now that the essence of the optimization scheme used has been presented, including a discussion of the linear programming subproblem that must be both constructed and solved at each step along the ascent procedure, we can restate the management problem in these terms

$$\min \{F[X^{(k+1)}] = \{\nabla C[X^{(k)}] + \nabla P[X^{(k)}] \cdot X^{(k+1)} + \gamma\} \tag{13}$$

$$\text{subject to } g_i(X) = 0 \qquad i = 1, \ldots, m \tag{2}$$

$$g_i(X) \geq 0 \qquad i = m + 1, \ldots, p \tag{3}$$

$$x_i \geq 0 \qquad i = 1, \ldots, n \tag{7}$$

$$x_j \leq \beta_j \qquad j = 1, \ldots, s \tag{14}$$

$$x_j \geq \alpha_j \qquad j = 1, \ldots, s \tag{15}$$

where γ is a constant and where β_j and α_j are, respectively, upper and lower bounds on the s discharge variables at the $(k+1)$th iteration. As mentioned previously, setting these bounds is similar to selecting a step size with other gradient methods of nonlinear programming. The efficiency of the optimization scheme depends directly on the selection of these bounds.

The linearized objective function equation (13) may be expressed in a slightly different form as

$$F = \left(\frac{\partial C(X)}{\partial X}\right) \cdot X + \left(\frac{\partial P(X)}{\partial X}\right) \cdot X + \gamma \tag{16}$$

In the REQM problem, $(\partial C(X)/\partial X)$ is a vector of linear cost coefficients associated with traditional production inputs—and residuals handling, modification, and disposal activities—and $[\partial P(X)/\partial X]$ is a vector of marginal penalties associated with the discharge of each residual.

Given that $R = h(X)$, equation (4), we see from equation (10) that

$$\frac{\partial P(X)}{\partial x_j} = \sum_{i=1}^{q} \frac{2}{r_i} (\max[(R_i - S_i), 0]) \frac{\partial R_i}{\partial x_j} \qquad j = 1, \ldots, n \qquad (17)$$

The term $2/r_i$ (max $[(R_i - S_i), 0]$) represents the slope dp_i/dR_i of the ith penalty function evaluated at the point R_i. The term $\partial R_i/\partial x_j$ represents the marginal response of the ith descriptor of the natural world (or ecosystem) to changes in the discharge of the jth residual. Equation (17) may be expressed more generally as,

$$\frac{\partial P(X)}{\partial x_j} = \sum_{i=1}^{q} \frac{dp_i}{dR_i} \cdot \frac{\partial R_i}{\partial x_j} \qquad j = 1, \ldots, n \qquad (18)$$

or in matrix notation as

$$\frac{\partial P(X)}{\partial X} = \left(\frac{\partial R}{\partial X}\right)^T \cdot \frac{dp}{dR} \qquad (19)$$

For linear environmental systems, $\partial R_i/\partial x_j$ is an element of the matrix of transfer coefficients A when the environmental functions are expressed linearly as

$$R = h(X) = A \cdot X \qquad (20)$$

Hence, for the case of linear environmental systems, the marginal penalties may be expressed in matrix notation as

$$\left(\frac{\partial P}{\partial X}\right) = A^T \cdot \left(\frac{dp}{dR}\right) \qquad (21)$$

where $(\partial P/\partial X)$ is a vector of marginal penalties, A is a matrix of environmental transfer coefficients, and (dp/dR) is a vector of slopes of the penalty functions evaluated at R.

For the case of nonlinear environmental models, the situation is similar except that evaluation of the environmental response matrix $(\partial R/\partial X)$ is somewhat more involved, and, in addition, because the response is nonlinear, it must be recomputed for each state of the natural world. Evaluation of the $(\partial R/\partial X)$ response matrix for the nonlinear aquatic ecosystem model not only involves taking derivatives of a system of implicit nonlinear algebraic equations, but it also involves using the chain rule for differentiating a series of interrelated functional forms.

One major problem with a large nonlinear programming problem of the type discussed here is that it may frequently be nonconvex. If this is so, then the procedure for solving it may fail to find the optimum optimorum and instead come to a halt at a lesser peak in the objective function.

Appendix 5-B
Natural Systems Models

In both the text and appendix 5-A, we have referred to models of natural environmental systems and the central role they play in analysis for REQM. In this appendix we present a brief discussion of the various types of such models and the procedures by which they are incorporated in the larger model discussed in appendix 5-A.

Natural systems models—air and water dispersion, sediment and nutrient runoff from land, chemical reaction, and biological systems—may be used to describe the impact on the environment of energy and material residuals discharged from the production and consumption activities included in a regional REQM model. Generally, natural systems models are used to specify steady state ambient concentrations of residuals and population sizes of species, at various points in space throughout the region of interest, given a set of values for such environmental variables as stream flow and velocity, groundwater flow and aquifer characteristics, slope and soil erodability, wind speed and direction, atmospheric stability, and atmospheric mixing depth.

Some environmental models are easier to deal with than others within a mathematical optimization framework. In general, this depends on the mathematical structure of the model. In terms of the complexity involved, it is useful to distinguish among four broad categories of models: (1) linear, explicit functions; (2) linear, implicit functions; (3) nonlinear, explicit functions; and (4) nonlinear, implicit functions.

Two natural systems submodels were actually used in conjunction with the regional REQM model discussed in appendix 5-A. The first, a linear atmospheric dispersion model, was used to predict ambient concentration levels of sulfur dioxide and suspended particulates throughout the region. The second, a nonlinear aquatic ecosystem model, was used to predict

AUTHORS' NOTE: This appendix is based on material prepared by Walter O. Spofford, Jr., and Robert A. Kelly.

ambient concentrations of various materials, and population sizes of certain species, throughout the water body involved, the Delaware Estuary. This model was developed at RFF by Robert Kelly specifically for the regional REQM study. These two models represent the extremes of complexity for inclusion within a REQM model framework. A discussion of each will raise some of the important issues and will reveal some of the problems involved and how they were handled.

Atmospheric Dispersion Model

Of the various atmospheric quality models that are now available, physical dispersion models are the most advanced.[1] Chemical reaction models, such as those dealing with photochemical smog, are in the early stages of application. The most successful modeling efforts to date have been associated with predicting both steady and nonsteady state distributions of sulfur dioxide, and particulates of 20 microns or more in diameter. Because of the availability of an existing air dispersion model, and the importance of these materials in the atmospheric environment of the region, ambient levels of sulfur dioxide and suspended particulates were selected to represent the air quality of the region.

The atmospheric model used in the REQM model is the dispersion model from the federal government's Air Quality Implementation Planning Program (IPP). This model is a point source Gaussian plume formulation. This formulation, which is based on mass continuity considerations, may be used to estimate ambient concentrations under deterministic, steady state conditions. For any given source-receptor pair, specified meteorologic conditions, and discharge rate of unity, this nonlinear equation reduces to a linear coefficient relating ambient concentrations with residuals discharge rates. This is characteristic of physical dispersion type models.[2]

The necessary inputs to the atmospheric dispersion model are: $x - y$ coordinates of all sources and receptors in the region; residuals discharge rates for each source—point and area; physical stack height, stack diam-

[1] For a description of these models, see, for example, H. Moses, *Mathematical Urban Air Pollution Models,* Report No. ANL/ES-RPY-001, Argonne, Ill., Argonne National Laboratory, April 1969.

[2] A model that reduces to a set of linear equations for steady state conditions is also available for analyzing the impact of organic residuals discharges on the dissolved oxygen regimes of streams. It is known as the Streeter-Phelps model. This type has frequently been used in water quality planning, but is not used in this study, having been replaced by the nonlinear aquatic ecosystem model.

eter, stack exit temperature, and stack exit velocity for each point source; a seasonal joint probability distribution for wind speed, wind direction, and atmospheric stability; mean seasonal temperature and pressure; and a mean atmospheric mixing depth for the period of interest. The output of this dispersion model represents arithmetic mean seasonal concentrations of sulfur dioxide and suspended particulates based on the probabilities of occurrence of a large number of discrete meteorological situations.

For a given set of meteorological conditions and physical parameters, the vector of mean seasonal concentrations of sulfur dioxide and particulates R may be expressed linearly, in matrix notation, as

$$R = AX + B \tag{1}$$

where X is a vector of sulfur dioxide and particulate discharge rates; A is a matrix of transfer coefficients which specify, for each source-receptor pair in the region, the contribution to ambient concentrations associated with a residuals discharge rate of unity; and B is a vector of background concentration levels. The matrix of transfer coefficients A is the output of the dispersion model.

The important thing to note from equation (1) is that the state of the natural world R is expressed directly in terms of linear, explicit algebraic functions. This particular mathematical form is relatively easy to deal with in an optimization framework. In fact, as is readily seen, equation (1) may be incorporated directly within the constraint set of a standard linear program when one of the management objectives is to constrain ambient concentrations of residuals.

As indicated in equation (16) and the associated text in appendix 5-A, one of the requirements of the optimization scheme used is the availability of an environmental response matrix, $(\partial R_i / \partial x_j), i = 1, \ldots, n, j = 1, \ldots, m$, where n is the number of designated receptor locations in the region and m is the total number of residuals discharges. This matrix may be obtained by differentiating equation (1) with respect to all the residuals discharges in the region, that is

$$\left(\frac{\partial R}{\partial X} \right) = A \tag{2}$$

or simply the matrix of transfer coefficients.

Before leaving this section, we should point out that not all atmospheric quality models are as easy to deal with as the physical dispersion models, which are expressed in linear, explicit analytical form. Chemical reaction models, such as for photochemical smog, would be significantly more difficult to handle within a mathematical optimization framework. The kinds of problems we would face with them are revealed in the discussion of a nonlinear aquatic ecosystem model which follows.

Aquatic Ecosystem Model

A variety of indicators are commonly used for describing the quality of a body of water. Among them are: bacterial counts, algal densities, taste, odor, pH, turbidity, suspended and dissolved solids, dissolved oxygen, temperature, and population sizes of certain plant and animal species. Because of the importance of dissolved oxygen to virtually all species of higher animals, the ease of measurement, and the relative ease with which it can be modeled for a river or estuary, its concentration has been, and still is, one of the most frequently used criteria for setting general water quality standards.

Streeter-Phelps-type dissolved oxygen models have been used for many years to predict water quality as a result of discharges of organic material (most notably, sanitary sewage). As we pointed out earlier, the steady state solution of such models results in a matrix of transfer coefficients which is exactly parallel in its use in the REQM model to that resulting from the steady state solution of the atmospheric dispersion model.

However, these models have three major deficiencies which the RFF modeling team felt warranted the development of a more sophisticated aquatic ecosystem model. First, there is interest in the dissolved oxygen level only insofar as it is an accurate indicator of such things as algal densities and the population sizes of certain species of fish. To the extent that these densities and populations can vary independently of dissolved oxygen concentrations, we need information about them independently if policies on water quality are to be established intelligently. Second, materials other than organic (for example, nutrients and toxics) are known to have significant effects on aquatic ecosystems. Consequently, these inputs should be explicitly included along with the organics to evaluate more fully the impacts on the aquatic environment of residuals discharges. Finally, systems ecologists feel that aquatic ecosystem models based on at least some biological (or ecological) theory that includes the mechanisms of feeding, growth, predation, excretion, death, and so forth, are more reliable for prediction purposes than the more empirically based linear models of the Streeter-Phelps variety, even for predicting dissolved oxygen levels.

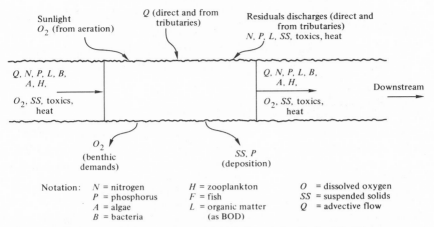

Figure 5-B-1. Inputs and outputs of a single reach: Delaware Estuary ecosystem model

The aquatic ecosystem model developed at **RFF** is based on a trophic level approach. The components of the ecosystem are grouped in classes (compartments) according to their functions, and each class is represented in the model by an endogenous, or state, variable. Eleven components (compartments) are designated in the model. The endogenous variables representing these eleven components are: nitrogen, phosphorus, turbidity (suspended solids), organic material, algae, bacteria, fish, zooplankton, dissolved oxygen, toxics, and heat (temperature). In addition, the following exogenous variables are considered: turnover rate (or advective estuary flow) and inputs (of the eleven chemical and biological materials above). Carbon is assumed not to be limiting and, hence, is not considered as either an endogenous or exogenous variable. Inputs to and outputs from a single reach of the model are illustrated in figure 5-B-1; internal material flows within a single reach are depicted in figure 5-B-2.

The inputs to the estuary model from the residuals discharges in the region are: organic material measured by its BOD_5, total nitrogen, phosphorus, phenols (toxics), and heat. The outputs are: densities of fish biomass, algal densities, and dissolved oxygen concentrations. The levels of these outputs are constrained, that is, environmental standards are imposed. In addition, concentrations of nitrogen, phosphorus, suspended solids, and organic material; temperature; and mass of bacteria and zooplankton are also available as by-product outputs of this model.

The time rate of change of material in each compartment is expressed in terms of the sum of the transfers among other compartments and

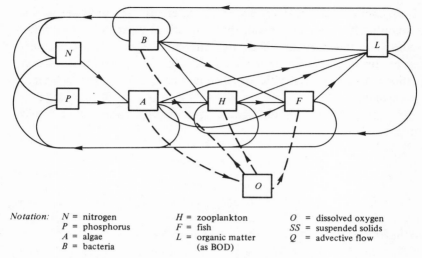

Notation: N = nitrogen H = zooplankton O = dissolved oxygen
          P = phosphorus F = fish SS = suspended solids
          A = algae L = organic matter Q = advective flow
          B = bacteria   (as BOD)

Note: The three remaining endogenous variables—heat (temperature), toxics, and suspended
          solids—are assumed to affect the rates of material transfers among the ecosystem com-
          ponents, or "compartments."

Figure 5-B-2. Internal materials flows within a single reach: Delaware Estuary ecosystem model. From: Walter O. Spofford, Jr., Clifford S. Russell, and Robert Kelly, "Operational Problems in Large-scale Residuals Management Models," in Edwin S. Mills, ed., *Economic Analysis of Environmental Problems* (New York, National Bureau of Economic Research and Resources for the Future, 1975) p. 185.

between adjacent sections of the estuary (since the material is distributed spatially as well as temporally). In order to ensure mass continuity of the materials considered, all materials entering and leaving a compartment are explicitly accounted for. By convention, oxygen is not included in the mass balances for species and materials, except, of course, for the oxygen balance itself. For the three nutrients—carbon (C), phosphorus (P), and nitrogen (N)—a mass balance is made on the individual chemical elements. For species, a mass balance is made on the total weight of carbon, phosphorus, and nitrogen. Fixed proportions of $C:P:N$ are assumed in the living components. The mathematical description of material transfers among compartments is based on the theoretical and empirical formulations given by Odum.[3]

[3] Howard T. Odum, "An Energy Circuit Language for Ecological and Social Systems: Its Physical Basis," in Bernard C. Patten, ed., *Systems Analysis and Simulation in Ecology,* Vol. II (New York, Academic Press, 1972).

Each compartment requires a separate differential equation to describe mass continuity, and, in general, these equations must be solved simultaneously. In this particular case, the differential equations are ordinary ones of the first order, nonlinear variety. In addition to these eleven equations, a set of similar differential equations is required for each reach of the estuary.

The general form of the differential equation set for the kth reach may be expressed as,

$$\left(\frac{dR}{dt}\right)^{(k)} = f[R(t)^{(k-1)}, R(t)^{(k)}, R(t)^{(k+1)}, X(t)^{(k)}] \tag{3}$$

where $(dR/dt)^{(k)}$ is a vector of time rates of change of the endogenous variables in the kth reach, $R^{(k)}$; and $X^{(k)}$ is a vector of residuals discharges into the kth reach.

Equation (3) is a set of ordinary nonlinear differential equations—one equation for each compartment and one set of compartmental equations for each estuary reach—that must be solved simultaneously. If one is interested in the transient (or nonsteady) states of the system, simulation techniques, that is, numerical integration (simulating over space and then time) furnishes a readily available means of solution. However, only the steady state solution was applied in the regional REQM model. There are two possibilities for determining steady state solutions, neither of which guarantees finding a stable point equilibrium: (1) simultaneous simulation of a nonlinear differential equation set; and (2) simultaneous solution of a set of nonlinear algebraic equations. If inputs to and outflows from each reach due to longitudinal diffusion are put aside, the system can be dealt with over time first, and then over space, starting with the uppermost reach and progressing systematically down the estuary.[4] In this case, equation (3) for the kth reach would reduce to

$$\left(\frac{dR}{dt}\right)^{(k)} = f[R(t)^{(k-1)}, R(t)^{(k)}, X(t)^{(k)}] \tag{4}$$

[4] Neglecting longitudinal diffusion, even in an estuary, is not as unreasonable as it first appears. Finite difference techniques for solving these differential equations introduce a numerical diffusion effect into the model. Inputs are immediately mixed in the volume, not because of any physical effects, but solely because of the numerical procedure. See Donald J. O'Connor and Robert V. Thomann, "Water Quality Models: Chemical, Physical, and Biological Constituents," in U.S. Environmental Protection Agency, *Estuarine Modeling: An Assessment* (Washington, D.C., GPO, February 1971) stock no. 5501-0129, chapter III, p. 138.

Now only the eleven compartmental equations within each reach must be solved simultaneously. The state of the system within a particular reach depends only upon the inputs from upstream $R(t)^{(k-1)}$ and the residuals discharges to the kth reach $X(t)^{(k)}$ both of which may now be treated as exogenous inputs. In addition, if the resulting steady state solution $R^{*(k)}$ is independent of the time paths of rates of inputs, $R(t)^{(k-1)}$ and $X(t)^{(k)}$, equation (4) reduces to

$$\left(\frac{dR}{dt}\right)^{(k)} = f[R^{*(k-1)}, X^{*(k)}, R(t)^{(k)}] \tag{5}$$

Usually, ecological models are solved by simulation. Simulation of the differential equation set [a set of equations similar to equations (3) through (5)] poses no particular problem, but the steady state solution, if one exists at all, may take considerable time.

Even when a steady state solution can be found, an additional problem is that there may be more than one stable point equilibrium. To investigate this problem, an experiment was run with the ecosystem model. A random number generator was used to provide a set of random starting points. Twenty-five random starts resulted in the same steady state solution.

At steady state, $(dR/dt) = 0$, and thus the differential equation set above, equation (5), reduces to a set of nonlinear algebraic equations of the following form

$$0 = f[R^{*(k-1)}, X^{*(k)}, R^{*(k)}] \tag{6}$$

The endogenous variables $R^{*(k)}$ are implicitly expressed in this formulation.

To include this nonlinear aquatic ecosystem model within the regional REQM model, in addition to determining a set of steady state values, it is necessary to evaluate the response throughout the ecosystem to changes in the rates of the residuals discharges. That is, it is necessary to know, for example, the effect on algae in reach seventeen of an additional BOD_5 load discharged into reach eight, and so forth. In contrast to the linear steady state models, this requirement results in a considerable number of additional computations, but this is the key to being able to use these complex ecosystem models within the optimization framework.

The response matrix we wish to compute may, as has already been pointed out above, be expressed in matrix notation as $(\partial R/\partial X)$, where R is a vector describing the state of the system throughout the entire length of the estuary, and X is a vector of residuals discharges throughout the region. Using equation (6) for each reach of the estuary, and the relationship,

$$Z^{(k)} = q^{(k)}R^{(k-1)} + X^{(k)} \tag{7}$$

where $Z^{(k)}$ is a vector of inputs to the kth reach; $R^{(k-1)}$ is a vector of concentrations of materials in reach $(k-1)$; $q^{(k)}$ is the estuary advective flow rate into the kth reach; and $X^{(k)}$ is a vector of residuals discharges to the kth reach, a section of the system response matrix may be computed according to

$$\frac{\partial R^{(i)}}{\partial X^{(j)}} = \frac{\partial R^{(i)}}{\partial Z^{(i)}} \cdot \frac{\partial Z^{(i)}}{\partial R^{(i-1)}} \cdot \frac{\partial R^{(i-1)}}{\partial Z^{(i-1)}} \cdots \frac{\partial R^{(j+1)}}{\partial Z^{(j+1)}} \cdot \frac{\partial Z^{(j+1)}}{\partial R^{(j)}} \cdot \frac{\partial R^{(j)}}{\partial Z^{(j)}} \cdot \frac{\partial Z^{(j)}}{\partial X^{(j)}}. \tag{8}$$

From equation (7) we note that

$$\frac{\partial Z^{(k)}}{\partial R^{(k-1)}} = q^{(k)}I \tag{9}$$

and

$$\frac{\partial Z^{(k)}}{\partial X^{(k)}} = I, \tag{10}$$

where I is the identity matrix. Thus, the $[\partial Z^{(k)}/\partial R^{(k-1)}]$ terms are known a priori and are exogenous parameters in the ecosystem model.

The other terms $[\partial R^{(k)}/\partial Z^{(k)}]$ are evaluated from equation (6) according to the rules for differentiating implicit functions. That is

$$\left(\frac{\partial f}{\partial R}\right) \cdot \left(\frac{\partial R}{\partial Z}\right) = -\left(\frac{\partial f}{\partial Z}\right)$$

or

$$\left(\frac{\partial R}{\partial Z}\right) \cdot \left(\frac{\partial f}{\partial R}\right)^{-1}\left(\frac{\partial f}{\partial Z}\right) \tag{11}$$

This operation involves the inversion of the Jacobian matrix $(\partial f/\partial R)$. In addition, because the system of equations is nonlinear, the Jacobian matrix $(\partial f/\partial R)$ must be recomputed for each state of the natural world.

It should be clear, then, from the above discussion, that the major problems associated with including environmental models within the management framework is one of computer time. Nonlinear representations of the natural world increase the complexity and the number of calculations necessary for each iteration, but they also increase the realism, predictive capability, and information output of the REQM model.

CHAPTER 6

Collective Choice and REQM Problems

To motivate the discussion of how political, that is, collective choice, considerations were embodied in the structure of the Russell-Spofford conceptual model, we established the fact that collective choices have to be made about environmental quality. This conclusion leads inevitably to difficult fundamental questions about the structure of the political institutions for making them. This centrally important topic will be discussed before describing a case study of regional REQM.

The Theory of Collective Choice

Kenneth Arrow pioneered the modern study of collective choice processes in his landmark book, *Social Choice and Individual Values*.[1] In it he laid down a set of properties that a desirable social choice mechanism could reasonably be expected to have. These properties, as he has recently restated them, are:

1. Collective Rationality: In any given set of individual preferences, the social preferences are derivable from the individual preferences.
2. Pareto Principle: If alternative A is preferred to alternative B by every single individual, then the social ordering ranks A above B.
3. Independence of Irrelevant Alternatives: The social choices made from any environment depend only on the preferences of individuals with respect to the alternatives in that environment.
4. Non-dictatorship: There is no individual whose preferences are automatically society's preferences, independent of the preferences of other individuals.[2]

Arrow analyzed voting situations rigorously with these conditions in mind and found that, in general, no mechanism could be devised that consistently met them all. This is his famous "impossibility theorem."

[1] Kenneth J. Arrow, *Social Choice and Individual Values,* 2nd ed. (New York, Wiley, 1963).
[2] Kenneth J. Arrow, "Public and Private Values," in S. Hook, ed., *Human Values and Economic Policy* (New York, New York University Press, 1967) pp. 3–21.

Despite much further examination of voting systems, this theorem still stands essentially intact for the kind of choice mechanisms Arrow examined.

Arrow also proved that a two-party system was able to meet all four conditions and pointed out that his Possibility Theorem for Two Alternatives was, in a sense, the logical foundation of the Anglo-American two-party system.[3] Edwin Haefele, the political theorist of the RFF modeling team, extended Arrow's work on two parties by devising rules for choice of the positions each party should take relative to individual preferences.[4] The essence of his case is that the two-party system can function in a way that brings out two positions on an issue which, when voted upon in a legislature, produce the same decision as would be generated if the voters could indulge in vote trading on that issue. It must be emphasized that Haefele found that representative government *could* operate as an ideal social choice mechanism, not that it does at present. This is analogous to saying, as we did earlier, that a competitive price structure is ideal under certain circumstances, but it is not the same thing as saying that present market prices and outputs are at the ideal point.

Essentially what Haefele's research has shown is that a political system based on representative government, majority rule, and a two-party system can, in principle, aggregate individual preferences in such a way that no further vote trading is possible.[5] This is the political analogue of the competitive equilibrium (Pareto efficiency) in economics.[6] Our existing political system resembles this idealized one in broad outline but diverges in many important details, such as the existence of the seniority system, regional representation, private financing of candidates and associated political payoffs, and the like—details which greatly attenuate its preference aggregation function. Moreover, as a society we have given very

[3] Arrow, *Social Choice*, p. 48.

[4] Edwin T. Haefele, "A Utility Theory of Representative Government," *American Economic Review* vol. 61, no. 3 (June 1971) pp. 350–67. Haefele's work is not an attack on Arrow's general theorem, as has erroneously been assumed by some, but rather builds on Arrow's theorem for two alternatives.

[5] Edwin T. Haefele, "Environmental Quality as a Problem of Social Choice," in Allen V. Kneese and Blair T. Bower, eds., *Environmental Quality Analysis: Theory and Method in the Social Sciences* (Baltimore, Johns Hopkins University Press for Resources for the Future, 1972).

[6] Strictly speaking, it also depends on some of the same kinds of assumptions: in market economics it is an equilibrium, given an income distribution and a set of resources and technologies; in public choice theory it is an equilibrium, given a set of issues and constitutional rules on voting. The latter political equilibrium will represent a point on a Pareto surface only if the issues themselves are Pareto-admissible, that is, could not be reformulated so as to make everyone better off.

little attention since the Constitution was written to the precise analysis of representative governmental institutions or to designing new ones to meet changing national needs.

If we are to have preferences consistently aggregated over the whole range of public goods issues, collective choice decisions must be made by properly representative legislative bodies encompassing the whole range of pertinent issues. Moreover, these issues must be formed within a two-party framework. Since they prominently include, but reach far beyond, questions of environmental quality, the most suitable mechanism for making these legislative or policy-type decisions is a government of general jurisdiction—not a specialized government agency.

Unfortunately, even if existing governments of general jurisdiction could be reformed so as to remove some of their most obvious divergencies from the utility model of representative governments, they do not suffice for all purposes. In our case, the central reason is that they seldom correspond in a reasonable way to the residuals "problemshed" boundaries. To take the case of air quality and solid residuals management, governments that have jurisdiction over extended metropolitan areas would be reasonably appropriate. However, this is not often the case. Instead, the numbers of separate governmental jurisdictions in a given metropolitan area sometimes total more than a hundred.

One of the main obstacles to the successful application of management science models to public problems is this Balkanization of government. Systems analysis aspires to analyze the interdependencies in systems. But the "decision makers" who are constantly being invoked by some economists and management scientists represent and are interested in only little pieces of what the scientist regards as the system. There is clearly a mismatch which surely helps to account for the frequent ineffectiveness of systems analysis. What we have raised is the question of optimal jurisdictions, and we shall discuss it further later.

Political Mechanisms for Making Collective Choices[7]

If one accepts that the preferences of individual citizens are to determine the proper mix between public and private goods and the types and

---

[7] The subject matter of the remainder of this chapter is treated in greater detail in Paul R. Portney, Jon C. Sonstelie, and Allen V. Kneese, "Environmental Quality, Household Migration, and Collective Choice" in Edwin T. Haefele, ed., *The Governance of Common Property Resources* (Baltimore, Johns Hopkins University Press for Resources for the Future, 1974).

amounts of public goods provided, then the essential attribute of any desirable collective decision-making mechanism is its ability to reflect the intensity as well as the direction of individual preferences. The value judgment that individual (or household) preferences are what counts also underlies the economic theory of resources allocation to which we have appealed so frequently. It is then axiomatic that we prefer those collective decision mechanisms which have the ability not only to express these intensities but also to arrive at collective decisions in such a way that more intensive preferences are, whenever possible, satisfied. It is against this criterion that we examine several different types of collective decision mechanisms.

We will claim that legislative vote trading could function effectively to aggregate preferences. Before indicating how a vote-trading model can be fitted into a regional REQM model, we examine some other mechanisms that are commonly used to make political decisions about public goods— primarily to justify our preference in principle for vote-trading outcomes as compared with those of various other collective decision-making mechanisms. We also consider the place of vote trading in a representative government more deeply—especially with respect to the important matter of the jurisdictional boundaries for institutions involved in making collective choices.[8]

Popular Referenda

Decision making by popular referendum is a method by which questions (most often about changes in expenditures or revenue raising) are placed on a ballot and put before the voters of a jurisdiction. The issue is usually decided by a simple majority of those voting.[9] The popular referendum

[8] In the remainder of this chapter we wish to examine how various collective decision-making mechanisms would function under ideal conditions. Accordingly, we assume that the preference function of an elected representative accurately reflects the preferences of his constituents. In practice, this would never be perfectly true because of imperfections in knowledge. Moreover, such imperfections in the political system as the representative being beholden to campaign contributors and accepting payoffs, distort the reflection of constituents' preferences in the representatives' decisions. Accordingly, political models such as those used here stand in somewhat the same relation to the political world as the competitive market models referred to in earlier chapters do to the actual economic world. They are idealized standards against which the performance of certain political structures in the real world can be measured.

[9] In some cases (some bond issues, for example), a 60 percent or two-thirds majority is required.

has seen its widest use for proposed increases in local property taxes to finance educational expenditures. However, it is not restricted to such questions, as witnessed by increasing reliance on referenda to settle questions of environmental importance, such as establishment of state jurisdiction over coastal zones, regulation of nuclear power plant construction, and the required use of returnable beverage containers.

The major problem with a popular referendum decision is that individuals and households can indicate no more than whether they favor the specific proposal or oppose it. That is, there is no way for them to indicate how strongly they approve or disapprove of the package. Those who would benefit greatly would vote "yes" as would those whose net gain would be small but positive. Conversely, those who would lose greatly from the proposal could only cast the same "no" vote as those who would bear small net losses. This inability to indicate intensity of preference is the major shortcoming of the referendum, because it renders nearly impossible the capture of potential mutual gains from trading.[10]

To see this more clearly, consider the case of a community which must reach a decision on two separate issues. One issue might be an increase in educational expenditures and the other an environmental improvement, both kinds of expenditures to be financed by increases in property taxes. Suppose that there are a number of houseowners who stand to gain significantly and directly from the approval of the educational expenditure increase via improved educational opportunities for their children, but that a majority of citizens will suffer slight income losses, say, because increases in their property taxes will outstrip the increases in the market values of their houses. Suppose also that a minority of citizens would benefit greatly from the environmental improvement but that more citizens would realize slight net losses from it.

If these two issues were resolved by popular referenda in the community in which individuals vote in a rational and self-interested manner, both the education and the environmental proposals could be defeated. Quite simply, more people oppose each proposal than favor it, and both issues are settled by simple majority. Notice, however, that those in a minority on each issue feel intensely about the issue (since they stand to make large gains), while those in the majority are only passively opposed. If agreements between voters could be made, then some voters who slightly oppose the improvement in the environment but who strongly favor the increase in education would be willing to support the former in exchange

[10] There is, however, some indication of intensity from the size of the turnout.

for the support on the education issue by citizens who are slightly opposed to it but who have a large stake in the environmental improvement.

It is this sort of agreement among voters that could make Pareto improvements possible (analogous to gains from trading in the market). Unfortunately, it is precisely this kind of arrangement that popular referendum voting makes impossible.[11] Moreover, such arrangements are most unlikely under a popular referendum mechanism because there are too many voters to facilitate the formation of majority coalitions; that is, the information requirements are much too large, and also there is no way under popular referendum to ensure that any kind of agreement voters might make would be kept. Enforceability is impossible with secret balloting, and there is an incentive for each voter to enter into agreements with other voters verbally, but to "cheat" when it comes time to cast the vote promised to another.

Multiple Representation

Multiple representation is the practice by which the citizens of a jurisdiction elect specific individuals to represent them on separate boards or commissions, each empowered to make decisions on a specific public good. For example, in some areas, boards of education make all decisions concerning educational expenditures and revenues; public safety commissions make similar decisions on police and fire protection matters; and public sanitation commissions decide questions of sewage treatment and garbage collection. Citizens are given the chance to elect members to these boards to represent their preferences, but the important feature is that each board oversees only one public good and, hence, each representative votes only on matters relevant to the board on which he sits.

As a result, a multiple representation collective decision mechanism suffers from the same defect that a popular referendum does. Namely, multiple representation does not allow voters (or their representative) to express intensities of preference on a set of independent issues.[12] There is no way for a school board representative, say, to indicate that his constitu-

---

[11] In principle, intensities could be registered in referendum voting by requiring the voter to register his vote on an intensity scale. But this would still not permit the direct tradeoffs against other issues possible with vote trading.

[12] The issues are not independent in the sense that normally all involve expenditures, and hence involve decisions on allocation of the resources in a "single pot." Consideration of all issues simultaneously is the only way in which their relative importance can be determined.

ency favors a certain proposal but only in a lukewarm fashion; he can only vote his constituents' preferences as he interprets them. Because he only has a vote on educational decisions, moreover, he cannot arrange a trade directly.[13] Nor can he form a coalition with a sanitation commission representative, for example, even though such a coalition based on intense minority interests might be desirable from the standpoint of many or all voters.

Multiple representation is a highly relevant issue for integrated residuals management because environmental quality issues are usually considered separately from other public goods issues, and even the various environmental media are often objects of separate political decision making.

The fact that multiple representation fails to provide a means by which voters or their representatives can bargain and coalesce over a number of issues in such a way that intense minority interests are sometimes satisfied makes it unattractive for collective decision-making purposes. This may be especially significant when considering environmental preservation versus commercial or industrial development, for example, because environmentalists often constitute a particularly intense minority in the real world. We turn our attention now to a collective decision mechanism that does facilitate the formation of agreements and, consequently, the representation of intense minority interests.

Geographic Representation

Geographic representation, in our sense, is a situation in which representatives are elected from small geographic units (wards or districts) to serve in a legislative body that considers all public goods issues simultaneously for the whole jurisdiction.

The primary advantage of geographical representation in a defined area such as a metropolitan region or river basin is that it facilitates the kinds of agreements necessary to secure the mutual gains from trading discussed above. Appendix 6-A at the end of this chapter develops the analytics of vote trading. But we can illustrate rather simply how the mechanism works in general. Suppose that there are two proposals before a simplified metropolitan government consisting of one representative from each of four constituent communities—A, B, C, and D. The first

---

[13] Although such trades may be arranged through informal power structures.

proposal provides for an increase in metropolitan-wide expenditures on transportation, and the second, for increased expenditures for improving some aspect of environmental quality that primarily benefits communities A and B. Let us assume that many of the residents of communities A and B are in favor of the environmental quality improvement but that they oppose increases in transportation expenditures. Assume that the former issue is of much greater concern to them than the latter, however. On the other hand, the residents of communities C and D are assumed to be opposed to the proposal for improving environmental quality (because it benefits them less than it costs them), and we assume—in order to make our point—that they are strongly in favor of increased transportation expenditures (such expenditures might, for example, put their houses within commuting distance of a major employment center and hence increase the values of those houses).

Using a method developed by Edwin Haefele and explained more fully in appendix 6-A, we can represent these preferences and intensities in a simple way that clearly points out the possible gains from the exchange of votes.

Table 20 indicates that the representative to the metropolitan government from community A opposes the transportation expenditure increase (as indicated by the N for "no") and that he is in favor of the environmental quality enhancement proposal (Y for "yes"). The subscripts indicate the relative intensities of his preferences, that is, environmental quality enhancement is more important to him (he would rather see the environmental issue pass than see the transportation issue fail). The same preferences are held by community B's representative. The representatives from C and D have opposite stands and intensities. If passage of any proposal requires the votes of three of the four representatives, both issues will clearly fail in the absence of agreements between representatives.

However, if the representative from A is willing to switch his "no" vote on the transportation issue (about which he is little concerned) to a "yes" vote, the representative from C, who intensely favors the transportation expenditure, would be willing to switch his vote on the environmental expenditure from no to yes. The relative intensities would then be shown as Y_2 on transportation for for A and Y_2 on environment for C, as shown in panel B of table 20. The result of such a trade would be three yes votes on each issue, and they would therefore both pass. Obviously, if unanimity were required, the representatives from B and D could in this instance arrange an identical trade. With both issues passing, most voters

TABLE 20.   Voter Preferences and Intensities

	Panel A before switching				Panel B after switching			
	Community				Community			
Issue	A	B	C	D	A	B	C	D
Transportation	N_2	N_2	Y_1	Y_1	Y_2	N_2	Y_1	Y_1
Environmental quality enhancement	Y_1	Y_1	N_2	N_2	Y_1	Y_1	Y_2	N_2

Note: N = no; *Y* = Yes.

would find that their intense preferences had been satisfied—something that would not have happened if separate referenda on the two issues had been employed or if an elected transportation board had resolved the one issue and an environmental quality commission the other. The advantage of general representative government (that is, of representative government responsible for all issues) is that it greatly increases the flow of information by reducing the number of decision makers on any question, enabling them to consider a wide variety of issues and various means of resolving problems or achieving goals. Moreover, they would be accomplishing this in a forum in which such agreements or vote trades could take place.

Although vote trading permits dealing with heterogeneity in a single political jurisdiction, it does not directly address the question of appropriate jurisdictional boundaries. Madison's implied definition for a political jurisdiction was a heterogeneous population having common problems. This probably had much to do with giving us territorial representation, as opposed to "interest" representation on a national basis, but it says little about how to carve up the territory.

Environmental quality issues illustrate the boundary problem quite clearly. The liquid, gaseous, and solid residuals problemsheds of an area will rarely coincide. Should these be three separate governmental agencies, even though the technology and economics of residuals management indicate the logic of having one integrated system? Whose views should be represented in such an integrated system, that is, the largest geographical set (usually the watershed)? To represent such a large unit will bring into the decision some people who are remote from the area, who care little about its problems, and who will propose solutions that force costs on other people.

General Purpose Representation

Haefele has suggested that one component of the boundary question could be resolved by electing a general purpose representative (GPR) at a district level, the districts being smaller than any government of general jurisdiction.[14] Such representatives would be members of all local and regional governmental bodies having jurisdiction over the district as measured, for example, by taxing authority or control over land use. (What is implied is a set of noncontiguous, often overlapping, functional areas.) The purposes of this building-block approach are twofold: (1) to enable the representative to have control over the whole range of local issues so that he could use his vote in one assembly as a lever in another, thus providing an opportunity for registering intensity of preferences; and (2) to enable governments of varying territorial jurisdiction to be assembled (and perhaps more important, disassembled) easily and conveniently with no upheaval in the basic political fabric of the area.[15] The suggestion has the added advantage of focusing local politics on one election so that citizen participation in the party structure and electioneering have more potential payoff. The representative, be it a man or a woman, is less vulnerable to special interest groups if he sits on the sewer board, zoning board, and the school board, as well as on a local government council, than he would be if he sits on only one board. The functional areas themselves could be run administratively under a professional manager type of structure.

It seems reasonable to assume that the general purpose representative system could also go a long way toward overcoming voter apathy in local elections, because all issues would be focused on the election of one man much as they are in the election of a governor, a senator or representative, or the president. Moreover, by focusing all issues in one election, the tendency for majority tyranny to emerge (in small populations such as the district) is mitigated. Intense preferences of minorities on one issue can be used to advantage in electoral politics, as well as in legislative politics.

Moreover, in the context of our present concern with REQM in a metropolitan area or region, the general purpose representative provides,

[14] Edwin T. Haefele, "General Purpose Representatives at the Local Level," *Public Administration Review* vol. 33, no. 2 (March-April 1973) pp. 177–179.

[15] For example, a district could opt into any local governmental body by being willing to be taxed by that body. It could not opt out without the permission of the governmental body concerned.

perhaps, a way out of the dilemma posed by the mismatch of jurisdictions and problems. A management agency whose reach encompasses the problem of, say, an expanded river basin commission could be put under the policy control of a representative body composed of the GPRs covering that region.

We recall that the major advantage of general purpose representation over geographic representation (say, for an entire river basin) is that it eliminates from the game players who have absolutely no direct interest in some of the issues. For example, in the case of geographic representation, a vote trader may have an interest in only one issue before a legislature but has the right to vote on many issues which the legislature must consider. In such a situation, he could secure passage of his desired issue by trading his "free" votes to other voters intensely interested in the issues to which those free votes may be applied. While the essence of vote trading is the exchange of votes on issues of little concern for votes on issues of great concern, the possession of such power on the part of a few representatives may lead to an overinvestment in the public goods which benefit their constituents.

General purpose representation eliminates this problem, because the representative of a particular jurisdiction does not have a vote on an issue unless his constituents derive benefits from, or bear costs of, the public good under consideration.[16] "Free money" is thus eliminated. Whether or not the problem would in fact be a serious one depends upon the specific geographic area under consideration. If many or most public goods affect all the residents of the region under consideration, the general purpose representative is very similar to the geographic representative. Since almost all individuals, and hence their representatives, have some interest in most of the problems, free money is not a serious problem. If, on the other hand, the region is so large or the public goods are of such a nature that each public good affects specific residents, then a system of geographic representation could be expected to make free money available to many representatives on certain issues. In these situations, general purpose representation might be appropriate, and we have seen that REQM issues illustrate these incongruities very well.

But there is a shortcoming to general purpose representation which must be recognized. In a metropolitan region governed by general purpose representation, it is conceivable—but not likely—that a certain area

[16] It must be recognized that it might sometimes be very hard to determine who is affected by what public goods.

might be affected by the provision of only one or two public goods. In such a case, that region would have virtually nothing to "give up" in order to collect votes on issues about which its residents were intensely concerned. Since general purpose representatives do not have votes on all public goods decisions, one could in fact expect less vote trading in such a system than there would be if public goods decisions were made by a metropolitan government where each representative had a vote on every decision.

To illustrate the point, suppose that a particular region is composed of six districts, as in table 21, the first four of which are contiguous with communities A, B, C, and D. Assume that each of the districts is a general purpose district and that all the citizens of the six districts are commonly affected by air quality. Suppose also that: only districts 3, 4, and 6 are included in the watershed; only districts 2 and 3 are affected by the natural public good; and education is provided by each of the four communities. We could then illustrate voting "rights" by the matrix shown in table 21. In this table, an X in the cell indicates that the district representative has a vote in the appropriate governmental agency. For example, all district representatives have a vote on the air quality board, only representatives from districts 3, 4, and 6 have votes on the water quality board, and so forth. The education votes of the first four districts are bracketed to indicate that the only trades possible on those issues are between district representatives 1 and 2 on the one hand and between 3 and 4 on the other hand, since a district representative can only vote on the education decisions of his own community.

In table 21, the representative from district 1 has only one possible trade, the representative from 2 has two possible trades, the representative from 3 has five possible trades, the representative from 4 has four possible trades, the representative from 5 has no possible trades, and the representative from district 6 has two possible trades. Since each trade is counted twice, there is a maximum of seven possible trades under general purpose representation, even though there are four issues and six potential traders. If a geographical government were formed to resolve the air, water, and natural public goods issues (education would not be included since we have defined it to affect only those inside each community), there would be a maximum of forty-five possible trades, more than seven times the maximum number possible under general purpose representation. General purpose representation, then, may be expected to cut down substantially on the number of vote trades, and this reduction will be

TABLE 21.   District General Purpose Representation

Issue	District					
	1	2	3	4	5	6
			Community			
	A	B	C	D		
Air	X	X	X	X	X	X
Water	—	—	X	X	—	X
Natural public good	—	X	X	—	—	—
Education	[X	X]	[X	X]	—	—

larger, the greater the number of issues affecting only a portion of the general purpose districts. Consequently, at least in principle, there is reason to prefer representative government arrangements based on geographical representation.

An Illustrative Application of Vote Trading

To illustrate how a collective choice mechanism can be included in a regional REQM model, the RFF model structure was applied to a simple prototype region, known as Didactica, containing twenty-five grids and only a few activities. This region is shown in figure 18. The simplicity of Didactica in comparison to the real case considered in chapter 7 makes it suitable for explaining how political elements were included in the overall model.

Providing for political as well as economic, technological, and ecological elements in the model was accomplished via the distribution of benefits and costs—a consideration central to the political process but, as noted previously, usually neglected or treated very simplistically in benefit–cost analysis. Geographical distribution across local political entities was made possible because the regional model is specific to a location in the sense that activities are assigned addresses in a grid. Accordingly, changes in environmental quality indicators, consumer price and tax increases, and employment for each different solution of a model can be associated with particular locations (grids). These types of impacts serve in lieu of damage functions used in the version of the model described in chapter 5 and its appendixes.

The first step in politicizing the Russell-Spofford model was to consider the distributional effects of environmental controls explicitly. To show

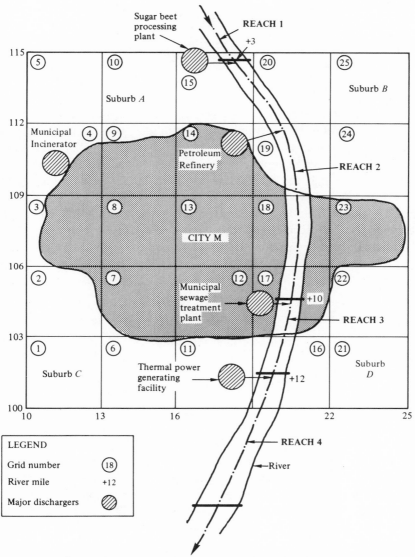

Figure 18. Didactica: locations of sources of residuals discharges and receptors of residuals, city *M* and surrounding suburbs *A, B, C,* and *D*. From: Clifford S. Russell, Walter O. Spofford, Jr., and Edwin T. Haefele, "The Management of the Quality of the Environment," in J. Rothenberg and Ian G. Heggie, eds., *The Management of Water Quality and the Environment* (New York, Macmillan/Halsted Press, 1974) p. 231 (reprinted by permission of Macmillan, London and Basingstoke).

how this was done, we characterize the Russell-Spofford model as the following linear programming problem:

$$\min c'X$$

$$\text{subject to} \begin{bmatrix} A_1 \\ A_2 \end{bmatrix} X \begin{array}{l} \geq b_1 \\ \leq b_2 \end{array}$$

where X is the vector of activity levels of the various production alternatives, modification and recirculation possibilities, and discharges; c is a vector of the unit costs of these activities; b_1 is a minimum "bill of goods" setting lower limits to production; b_2 is a set of ambient environmental quality constraints, for example, upper limits on SO_2 in all grids, upper limits on the dissolved oxygen deficit in all river reaches; A_1 is the technology matrix including residuals modification technology; and A_2 is composed of the several environmental-model transfer matrices. Thus, A_2 might be further partitioned as follows:

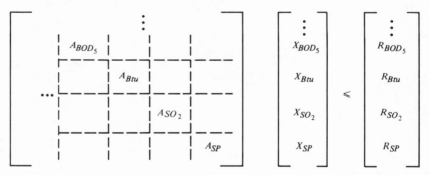

The right-hand side b_2 of course would be similarly partitioned, as we show. This model generates the cost of simultaneously meeting a certain level of production and ambient environmental quality standards at least cost. It is a relatively simple matter to include distributional information in this general model. All that is needed is to add a third part to the constraint matrix and a third set of values to the right-hand side. Thus, the problem becomes:

$$\min c'X$$

$$\text{subject to} \begin{bmatrix} A_1 \\ A_2 \\ A_3 \end{bmatrix} X \begin{array}{l} \geq b_1 \\ \leq b_2 \\ \leq b_3 \end{array}$$

TABLE 22.   Constraints on the Distribution of Costs Among Political Jurisdictions

Constraint on	Notes and description
1. Percentage increase in cost of electricity	Based on extra costs involved in adding cooling towers, precipitators, etc., compared to basic production process with no residuals control. Applies uniformly throughout region.
2. Percentage increase in costs of domestic liquid and solid residuals collection, modification, and disposal in city M and in suburbs A, B, C, and D	For city: based on incremental costs of going to secondary or tertiary treatment, of added sludge disposal, of cleaning incinerator stack gases, and of using more efficient landfill methods. For suburbs: costs of using more efficient (higher compaction) landfill methods and/or transporting to more distant sites. Also share of increased incineration costs if incinerator is used by suburb.
3. Unemployment at sugar beet processing plant and petroleum refinery	Sugar beet: based on degree of production cutback only. Petroleum: based on process changes as well as total production level.
4. Increase in annual heating cost of average household implied by burning lower sulfur fuel in city and each suburb	Based on following price differential: High sulfur DFO, \$4.35/Barrel (base case) Medium sulfur DFO, \$4.54/Barrel Low sulfur DFO, \$4.70/Barrel Annual figure obtained by blowing up average Nov./Dec.-day by annual average degree-day total for Philadelphia region.

Note: DFO = distillate fuel oil.

Source: Russell, Spofford, and Haefele, "The Management of the Quality of the Environment," p. 245 (reprinted by permission of Macmillan, London and Basingstoke).

where A_3 is structured to allow the product $A_3 \cdot X$ to represent costs over and above the *status quo* case incurred by various activities, that is, industrial plants, political jurisdictions, or other groups. Thus, one row of A_3 might represent increased costs of electricity to the customers of the regional utilities; another might be increased unemployment at a particular industrial plant implied by any cutbacks of production chosen to achieve overall production and environmental quality goals at least cost. b_3 is a vector of acceptable levels for these costs, for example, the largest acceptable increase in electricity costs or unemployment.

The distributional constraints actually used are summarized in table 22. The technique of their insertion in the model is straightforward and is shown schematically in figure 19. For example, to deal with the electricity cost constraint, a row called XTRCOSTE is added to the linear constraint

Row names (constraints)	Column names (activities)			Right-hand side
	Electricity production vectors EP_1 , ... , EP_n	Power import	INCELCOST	
⋮ XTRCOSTE ULXCSTE ⋮	$+ C_1$, ... , $+ C_n$	$+ C_I$	$- C_B$ $+ 1$	$=$ 0 $\leqslant$ (or 1.05, or 1.2)

Figure 19. Distributional constraints: increase in the cost of electricity. From: Russell, Spofford, and Haefele, "The Management of the Quality of the Environment," p. 245 (reprinted by permission of Macmillan, London and Basingstoke).

set. Each electricity production vector EP_j including the base process EP_1 has a positive entry C_i in that row giving the cost per megawatt hour of using it. (This includes the energy importation activity, EP_I, which is assumed to be more costly than domestic production.) Then, an activity that measures the percentage increase in cost, *assuming* a constant level of energy use, is added. This column is called INCELCOST. This vector has, as a negative entry in the XTRCOSTE row, a number equal to the total generation cost in the base case. In the simplest case this would be $EP_1 \cdot C_1 = C_B$. The sum across the row is constrained to be zero.

In Didactica a variety of opinions, desires, and dislikes related to the residuals aspects of environmental quality is assumed, as well as a wide variation in the willingness and ability to pay for improvements in environmental quality. To capture some of this complexity in the environmental issue, a preference vector that combines an ordinal ranking and a yes–no vote on a given set of issues is used. (See appendix 6-A for a more detailed explanation of preference vectors.) In the legislative model for Didactica, the issues consist of (1) four environmental quality indicators: dissolved oxygen deficit (in parts per million [ppm]), change in water temperature (in degrees F), suspended solids in micrograms per cubic meter ($\mu g/m^3$), and sulfur dioxide ($\mu g/m^3$) and (2) four measures of increased cost resulting from improvements in environmental quality, the latter indicated in table 22.

In Didactica, arbitrary preference vectors (reflecting differences in tastes, incomes, and the like) were specified for each of the twenty-five grids into which the area of Didactica was divided. (In the case study report in chapter 7, the preference vector for each subarea of the region was based on a detailed examination of the socioeconomic characteristics of the subarea.) These vectors have arbitrary upper limits imposed on each of the eight issues previously specified. Thus, the vector of upper limits for grid 1 is:[17]

Grid 1

DOD	3.0	parts per million in reach 4
ΔT	5.0	degrees F in reach 4
SP	50.0	$\mu g/m^3$ in grid 1
SO_2	20.0	$\mu g/m^3$ in grid 1
Tax	1%	increase of 1 percent in suburb *C*
Unemployment	10%	increase of 10 percent at the sugar beet processing plant
Household elec.	50%	increase of 50 percent to each household in every grid
Household heat	$20	increase in dollars per year to each household in suburb *C*

If grid 1 only were the decision maker, these upper limits would be the constraints grid 1 would put on the solution of the regional model. Since grid 1 is only one of twenty-five grids in Didactica, a social choice process is constructed to allow the upper limits vectors of all twenty-five grids to be expressed. Preference vectors, one for each grid, are designed to do that, that is, to show the relative ranking of the individual measures. For example, grid 1's reaction (based on its preference vector) to run A (no constraints) is displayed in table 23.

In this case, the numerical subscripts in the preference vector indicate the ordinal ranking of three issues. Thus the rankings are, by assumption: an upper limit of 5°F heat rise in reach 4 of the river is of first importance in grid 1; an upper limit on SO_2 of 20 $\mu g/m^3$ is second in importance; and an upper limit of 10 percent on unemployment at the sugar beet processing plant is third in importance to grid 1. (For ease of computation, not all issues are ranked; elements without subscripts are assumed to be all of equal importance but less important than any subscripted element.) Our

---

[17] *DOD* = dissolved oxygen deficit; ΔT = change in temperature in degrees Fahrenheit; *SP* = suspended particulates; SO_2 = sulfur dioxide; Household Elec. = household electricity cost; Household Heat = household heating cost.

TABLE 23.   Grid 1: Preference Vector

	Grid 1 upper limits	Results of run A for grid 1	Grid 1 preference vector
			(Vote)
DOD	3.0	5.5	N
ΔT	5.0	9.9	N_1
SP	50.0	17.2	Y
SO$_2$	20.0	32.7	N_2
Tax	1%	0.0	Y
Unemployment	10%	0.0	Y_3
H. elec.	50%	0.0	Y
H. heat	$20	0.0	Y

Note: DOD = dissolved oxygen deficit; ΔT = change in temperature; *SP* = suspended particulates; SO$_2$ = sulfur dioxide; H. elec. = household electricity cost; H. heat = household heating cost; *N* = no; *Y* = yes.

Source: Russell, Spofford, and Haefele, "The Management of the Quality of the Environment," p. 253 (reprinted by permission of Macmillan, London and Basingstoke).

ordinal ranking is based on the assumed socioeconomic characteristics for the particular grid.

Grid 1's preferences are summarized in table 23, which shows that the citizens of grid 1 are dissatisfied with an unconstrained solution (no constraints on present production and consumption activities in Didactica) for three out of four environmental quality indicators, while they are satisfied with the present tax and utility burdens on grid 1. Since grid 1 ranks the environmental quality issues above the financial ones, however, some additional financial burden would be accepted if necessary to achieve acceptable levels of water and air quality.

All other grids were assigned upper limits vectors and ordinal ranks for three or more issues. Using those vectors, all twenty-five preference vectors in response to the results of run A on each of the measures (as shown in the middle column of table 23 for grid 1) can be displayed. These are shown in table 24. The number of *Y*-votes on each row are tallied in the far right column.

This table shows unanimous approval on all financial issues but much disapproval of the present quality of the air and water. Assuming the preference vectors are such that not all can be met simultaneously, the stage is set for some sort of social choice process to be invoked. Because the number of possible "solutions" (that is, technically feasible alternatives) may be almost infinite, the social choice process cannot be simply a blind groping for a solution acceptable to some given percentage of grids, for such a process would prove to be inefficient and the solution ambiguous

TABLE 24. Vote Matrix, Run A (Unconstrained)

Issue	\multicolumn{25}{c}{Grid}																									Talley of Y votes
	1	2	3	4	5	6	7	8	9	10	11	12	13	14	15	16	17	18	19	20	21	22	23	24	25	
DOD	N	N	Y	Y	Y	N	N	N	N	Y	N	N_5	Y	N	Y_3	N	N	N_1	N	N	N	Y_1	Y_1	N_2	Y	11
ΔT	N_1	N	N	N	Y	N_3	N	N	N	Y	N_3	N_4	N	N	Y	N_3	N	Y	Y	Y	N_2	N_2	Y	N_4	N	3
SP	Y	Y	Y	Y	Y	Y_2	Y_2	Y	Y	Y_2	Y_1	Y_2	Y_1	Y	Y_2	Y_2	Y_2	Y	Y	Y_1	Y_3	Y	Y	Y	Y_1	25
SO_2	N_2	N_1	N	Y	N	Y	N_1	N	N	N	N_2	N_1	N_3	Y	Y_1	N_1	N_1	N	Y_2	N	N_1	N_4	N	Y	Y_1	6
Tax	Y	Y	Y_1	Y	Y	Y	Y_3	Y_1	Y_2	Y	Y	Y_3	Y_2	Y	Y	Y	Y_3	Y_2	Y_2	Y_2	Y	Y_3	Y_2	Y_3	Y	25
Unemp.	Y_3	Y_2	Y_2	Y_1	Y_1	Y	Y_4	Y	Y_1	Y_1	Y	Y	Y	Y_1	Y_4	Y	Y	Y	Y_1	Y_3	Y	Y	Y_3	Y_1	Y	25
H. elec.	Y	Y	Y	Y	Y_2	Y	Y	Y_2	Y_3	Y_3	Y	Y	Y	Y_2	Y	Y	Y	Y	Y_3	Y	Y	Y	Y	Y	Y	25
H. heat	Y	Y_3	Y_3	Y	Y_3	Y	Y	Y	Y	Y	Y	Y	Y	Y	Y	Y	Y	Y	Y	Y	Y	Y	Y	Y	Y_2	25
Package[a]			✓	✓	✓			✓	✓	✓			✓	✓	✓				✓	✓			✓		✓	13

Note: DOD = dissolved oxygen deficit. ΔT = change in temperature. SP = suspended particulates. SO_2 = sulfur dioxide. H. elec. = household electricity cost. H. heat = household heating cost. N = no. Y = yes.

Source: Russell, Spofford, and Haefele, "The Management of the Quality of the Environment," p. 255 (reprinted by permission of Macmillan, London and Basingstoke).

[a] Where a grid's first and second ranked measures are voted Y.

at best. The method used is based on a vote-trading algorithm. The essence of vote trading is giving up on one issue to gain another issue you value more. The basic idea of the vote-trading algorithm is to add constraints to the unconstrained run (run A, table 24), such that N-votes are converted to Y-votes in some efficient, nonbiased way. Vote trading is efficient for this purpose because it focuses attention on high-ranked N-votes (these are the upper limits violations of most concern to those grids). They are the upper limits that the grids want most to be put in as constraints on the regional model. However, with vote trading, such a constraint can be put in only if the grid that wishes to put it in will accept, that is, also allow to be added as a constraint, another upper limit on another issue that is desired by another grid. Constraints are put into the solution, therefore, in pairs.

Because vote trading is explained as giving something up for something of higher value, what is it that each grid gives up by this trade? To illustrate what is given up, a vote trade is picked from table 24.

	Grid 13 upper limits $\rightarrow$	Grid 13 results from run A $\rightarrow$	Grid 13 preference vector	Grid 24 preference vector $\leftarrow$	Grid 24 results from run A $\leftarrow$	Grid 24 upper limits
DOD	3.5 (Reach 2)	3.4	Y	N_2	3.4	3.0 (Reach 2)
SO₂	80	103	N_3	Y	38	40

The result of the vote trade shown is to put in, as constraints, grid 24's upper limit of $DOD \leq 3$ on reach 2 of the river and grid 13's upper limit of $SO_2 \leq 80$ in grid 13. What grid 13 gives up is a 3.4 DOD outcome, an outcome with which it is happy since it is below its own limit of 3.5. Grid 24 gives up an SO_2 outcome of 38 in its grid, an outcome with which it is happy.

The reasoning underlying this trade is as follows: grid 13 says, I do not know what the effect will be of putting a tighter limit on DOD in reach 2 of the river, but I am willing to do it if I can get my SO_2 concentration down at least to my upper limit. Grid 24 says, I want reach 2 of the river cleaned up more than it is now, and I'm willing to support an upper limit on grid 13's SO_2 concentration if I can get support for a cleaner river. Neither knows what effects these constraints will have on other things of concern to them, for example, their tax bills.

In the algorithm, if other upper limits are violated on a subsequent run, the grids are allowed to vote N on those rows and to try to trade on those issues if they can find a partner to trade with. In this way, higher ranked limits are given priority but lower ranked limits are allowed into the solution at a later stage (they may or may not be feasible at that time).

One more rule must be specified before proceeding with the vote-trading algorithm. It is assumed that if a grid's first- and second-ranked measures are voted Y, then the grid is in favor of the solution; if they are not, then the grid votes N on the solution. These "package" votes are tallied in the bottom row of the trading tables (in table 24, there would be thirteen Y-votes for run A, the unconstrained case). Since thirteen is a majority of the twenty-five grids, it might be thought that the *status quo* (run A) is acceptable. However, the use of the vote-trading algorithm reveals that this majority is not a dominant majority.

The process is started by solving the regional model a second time with the two constraints from the trade between grid 13 and grid 24, which was the only trade in run A. The new solution generates a new vote matrix and hence new trading opportunities. The vote matrix resulting from the new solution is shown in table 25. The results are easily summarized by comparing the two tables: we have improved the oxygen in the river (eighteen votes for instead of eleven), lowered the $\triangle T$ also (eight for instead of three), improved SO_2 substantially (twenty-two for instead of six), lost support on the tax rate (seventeen rather than twenty-five), and still have unanimous satisfaction with the levels of SP, electricity, and heating bills. What we have lost is our unanimous vote on the unemployment rate, which has jumped to 12.5 percent and only seven grids will accept that. Moreover, for the whole package we have only nine Y-votes instead of thirteen.

As would seem likely, however, a number of trades are now possible, and when these are completed, a new run of the regional model can be made, adding pairs of constraints to the constraints already in. The legislative model proceeds in this fashion until there are no further trades possible. In this case this occurs after three iterations, with the result shown in table 26.

The value of the objective function has gone from $89,300 in the unconstrained solution to $85,600 in the last solution, indicating reduction in economic activity. The small number of Ns left in the matrix indicates that the initial upper limits preferences were not very demanding

TABLE 25. Vote Matrix After First Trade

Issue	1	2	3	4	5	6	7	8	9	10	11	12	13	14	15	16	17	18	19	20	21	22	23	24	25	Tally
															Grid											
DOD	N	N	Y	Y	Y	N	Y	Y	Y	Y	N	Y_5	Y	Y	Y_3	N	N	Y_1	Y	Y	N	Y_1	Y_1	Y_2	Y	18
ΔT	N_1	N	N	N	Y	N_3	N	N	N	Y	N_3	Y_4	N	N	Y	Y_3	N	N	N	N	N_2	N_2	Y	Y_4	Y	8
SP	Y	Y	Y	Y	Y	Y_2	Y_2	Y	Y	Y_2	Y_1	Y_2	Y_1	Y	Y_2	Y_2	Y_3	Y	Y	Y_1	Y_3	Y	Y	Y	Y_1	25
SO$_2$	N_2	N_1	Y	Y	Y	N_1	Y_1	Y_1	Y	Y	Y_2	Y_1	Y_3	Y	Y_1	Y_1	Y_1	Y	Y_2	Y_2	Y_1	Y_4	Y_2	Y	Y_1	22
Tax	Y	Y	N_1	Y_2	Y	Y	N_3	N_1	N_2	Y	Y	N_3	N_2	N_1	N_4	Y	N_3	N_2	Y_2	Y	Y	Y_3	Y_2	Y_3	Y_1	17
Unemp.	N_3	N_2	N_2	N_1	N_1	Y	N_4	Y	N_1	N_1	Y	N	N	N_1	N_4	Y	N_3	Y	N_1	N_3	N	N	N_3	N_1	Y	7
H. elec.	Y	Y	Y	Y	Y_2	Y	Y	Y	Y_3	Y_3	Y	N	Y	Y_3	Y	Y	Y	Y_3	Y_3	Y	Y	Y	N_3	Y	Y	25
H. heat	Y	Y_3	Y_3	Y	Y_3	Y	Y	Y	Y	Y	Y	Y	Y	Y	Y	Y	Y	Y	Y	Y	Y	Y	Y	Y	Y_2	25
Package	N	N	N	N	N	Y	Y	N	N	N	Y	Y	N	N	Y	Y	Y	N	N	Y	N	N	Y	N	Y	9

Note: DOD = dissolved oxygen deficit. ΔT = change in temperature. *SP* = suspended particulates. *SO$_2$* = sulfur dioxide. H. elec. = household electricity cost. H. heat = household heating cost. *N* = no. *Y* = yes.

Source: Russell, Spofford, and Haefele, "The Management of the Quality of the Environment," p. 260 (reprinted by permission of Macmillan, London and Basingstoke).

TABLE 26. Vote Matrix, No Further Trades Possible

Issue	Grid																									Tally
	1	2	3	4	5	6	7	8	9	10	11	12	13	14	15	16	17	18	19	20	21	22	23	24	25	
DOD	N	N	Y	Y	Y	N	Y	Y	Y	Y	N	Y_5	Y	Y	Y_3	N	N	N_1	Y	Y	N	Y_1	Y_1	Y_2	Y	18
ΔT	Y_1	Y	Y	Y	Y	Y_3	Y	Y	Y	Y	Y_3	Y_4	Y	Y	Y	Y_3	Y	Y	Y	Y	Y_2	Y_2	Y	Y_4	Y	25
SP	Y	Y	Y	Y	Y	Y_2	Y_2	Y	Y	Y_2	Y_1	Y_2	Y_1	Y	Y_2	Y_2	Y_2	Y	Y	Y_1	Y_3	Y	Y	Y	Y_2	25
SO_2	Y_2	Y_1	Y_1	Y	Y	Y_1	Y_1	Y	Y	Y	Y	Y_1	Y_3	Y	Y_1	Y_1	Y_1	Y	Y	Y_1	Y_1	N_3	Y_2	Y_3	Y_2	25
Tax	Y	Y	Y_1	Y_2	Y	Y_3	Y_3	Y_1	Y_2	Y	Y	Y_3	Y_2	Y_1	Y_4	Y_3	Y_3	Y_3	Y_2	Y_2	Y	Y	Y_3	Y_1	Y	24
Unemp.	Y_3	Y_2	Y_2	Y_1	Y_1	Y_4	Y_4	Y	Y_1	Y_1	Y	Y	Y	Y_1	Y	Y	Y	Y	Y_1	Y_3	Y	Y	Y_3	Y	Y	25
H. elec.	Y	Y	Y	Y	Y_2	Y	Y	Y_2	Y_3	Y_3	N	Y	Y	Y_2	Y	Y	Y	Y_3	Y_3	Y	Y	Y	Y	Y	Y	25
H. heat	Y	Y_3	Y_3	Y	Y_3	N	Y	Y	Y	Y	N	Y	Y	Y	Y	Y	Y	Y	Y	Y	Y	Y	Y	Y	Y_2	23
Package	Y	Y	Y	Y	Y	Y	Y	Y	Y	Y	Y	Y	Y	Y	Y	Y	Y	Y	Y	Y	Y	Y	Y	Y	Y	25

Note: DOD = dissolved oxygen deficit. ΔT = change in temperature. *SP* = suspended particulates. SO_2 = sulfur dioxide. H. elec. = household electricity cost. H. heat = household heating cost. *N* = no. *Y* = yes.

Source: Russell, Spofford, and Haefele, "The Management of the Quality of the Environment," p. 260 (reprinted by permission of Macmillan, London and Basingstoke).

or in conflict. The initial starting point and the rules for adding constraints did not, however, prevent arrival at the point at which all twenty-five grids would accept the package. Had the original majority been accepted, it would have clearly been at a Pareto-inferior solution. Tests were performed on the vote-trading algorithm by altering the upper limit vectors of each grid in various ways. In each case the algorithm found the dominant solution.

The preceding discussion assumed that a legislature existed which allowed the preferences of all twenty-five grids to be represented in the decision process. Rarely does such a mechanism exist in a region or in a metropolitan area. (An exception is in the Metropolitan Council of Minneapolis-St. Paul, organized from fourteen equal population districts.) A typical pattern in an American metropolitan area is a council of governments in which each municipality is represented. This pattern of preference aggregation can be duplicated in the model, and the usefulness of the model for testing alternative governmental structures can be illustrated by working with the city and four suburbs as follows:

Suburb A = grids 5, 10, 15
Suburb B = grids 19, 20, 24, 25
Suburb C = grids 1, 2, 6, 11
Suburb D = grids 16, 21, 22
City M = grids 3, 4, 7, 8, 9, 12, 13, 14, 17, 18, 23.

The preference vectors (assuming the original slack grid upper limits and ordinal rankings) are aggregated on some basis (say, simple majority) and the resulting vector used as representative of the municipality. For example, starting from the run A case (table 24), and combining grids 5, 10, and 15 yields the results shown in table 27.

The process of aggregating by municipality has masked the trading possibilities inherent in the grid-voting vectors. However, the typical council of governments aggregation correctly reflects that the unconstrained solution has majority support. But from the earlier vote-trading solutions it is known that the unconstrained case is not a dominant solution. Moreover, councils of governments often must operate on a consensus basis. The council of government staff could point out, on the basis of the resulting pattern of air quality, that putting an SO_2 upper limit on grid 13 of, say, 80 $\mu g/m^3$ would probably result in making more grids satisfied with the SO_2 level. Likewise, a limit of $\Delta T \leq 4°F$ in river reaches

TABLE 27. City-Suburban Vote Matrix Based on Run A

Issue	Municipality					Tally
	A	B	C	D	M	
DOD	Y	N	N	N	Y	2
ΔT	Y	N	N_3	N_2	N	1
SP	Y_2	Y_3	Y_2	Y_3	Y_4	5
SO_2	N	Y	N_1	N_1	N_3	1
Tax	Y_1	Y_2	Y	Y_4	Y_2	5
Unemployment	Y_3	Y_1	Y_4	Y	Y_1	5
H. elec.	Y	Y	Y	Y	Y_5	5
H. heat	Y	Y	Y_5	Y	Y	5
Package	Y	Y	N	N	Y	3

Note: DOD = dissolved oxygen deficit. ΔT = change in temperature. *SP* = suspended particulates, SO_2 = sulfur dioxide. H. elec. = household electricity cost. H. heat = household heating cost. N = no. Y = yes.

Source: Russell, Spofford, and Haefele, "The Management of the Quality of the Environment," p. 264 (reprinted by permission of Macmillan, London and Basingstoke).

TABLE 28. City-Suburban Vote Matrix Based on Run A Without Trades

Issue	Municipality					Tally
	A	B	C	D	M	
DOD	Y	N	N	N	Y	2
ΔT	Y	Y	Y_3	Y_2	Y	5
SP	Y_2	Y_3	Y_2	Y_3	Y_4	5
SO_2	Y_3	Y	N_1	Y_1	Y_2	4
Tax	Y	Y_2	Y	Y	Y_3	5
Unemployment	Y_1	Y_1	Y	Y	Y_1	5
H. elec.	Y	Y	Y	Y	Y	5
H. heat	Y	Y	Y	Y	Y	5
Package	Y	Y	N	Y	Y	4

Note: DOD = dissolved oxygen deficit. ΔT = change in temperature. *SP* = suspended particulates. SO_2 = sulfur dioxide. H. elec. = household electricity cost. H. heat = household heating cost. N = no. Y = yes.

Source: Russell, Spofford, and Haefele, "The Management of the Quality of the Environment," p. 264 (reprinted by permission of Macmillan, London and Basingstoke).

2 and 4 could be expected to improve the heat situation in the river. In the absence of trades, the council of governments might well take this advice. Setting these constraints and aggregating again by municipality results in the voting matrix shown in table 28.

There are now four municipal jurisdictions in favor of the package (although on a grid-by-grid count, five grids are still opposed). If a two-thirds majority were the decision rule used by the council of governments, this would be the final solution (there again being no trades possible), a solution Pareto-inferior to the grid-trading solution to the same problem. However, there is still no unanimity; therefore constraints continue to be added, on advice of the staff, to ease the pain in suburb C, the only N package vote remaining. The three easiest to meet SO_2 limits in suburb C were then added to the regional model. Adding these constraints results in adding three more Y grids (total 22), and when the grid-voting sections are aggregated again by municipality, they reveal that a unanimous Y vote has been achieved. All five municipalities favor the package, although only eleven grids favor the DOD level and three grids disapprove of the package. This solution is also Pareto-inferior to the earlier solution. The three grids that disapprove of the solution are, unfortunately for them, each in a different municipality. This analysis shows that councils of governments do not aggregate preferences as effectively as does vote trading by smaller subareas represented by the grids.

Concluding Comment

The above analysis of a hypothetical region demonstrates how a political dimension can be explicitly incorporated into an economic-technologic-ecologic analysis of REQM in a region. The multiple components of the preference vectors for each subarea of a region set the stage for a social choice process to come into play. The analytical process for aggregating preferences into social choices takes account of intensities of preferences as well as numbers for and against any issue. When preference vectors reflecting these intensities are combined, they reveal opportunities for vote trading, and vote trading provides the key to arriving at stable social choices.

The politicized regional REQM model provides an effective instrument for analyzing the results of alternative REQM options in terms of levels of ambient environmental quality and distributions of benefits and costs. It permits determining which patterns most closely fulfill the sets of preferences for alternative outcomes. This is a very useful characteristic because —to reiterate—seldom, if ever, can all relevant damage functions be defined in monetary terms. The politicized model also enables us to analyze the effects on decisions of alternative patterns of representative government. Given the fact that governmental structures, rather than technology as such, are very often the key to achieving improved ambient environmental quality, this attribute is important.

Appendix 6-A
Collective Choice and Representative Government

In chapters 5 and 6 we have established the need in REQM for a theory of collective choice through government institutions. This area of theorizing is in one sense very new and, as we note below, in another sense, very old. In this appendix we delve a little more deeply into the theory to show more clearly the importance of vote trading and the importance of representative government as a vehicle for achieving it.

One line of theoretical development views government as being a quasimarket in the sense that it simply registers and transmits individual preferences, on the basis of the distribution of votes. The prototype model of this sort would be based on the assumption that each public issue is resolved by a referendum universally participated in by fully informed voters. The "market type" formulation of the political process has frequently been criticized by those who note that most political decisions are and must be made through representatives. The market type of model has, however, given rise to a result which has been discussed almost endlessly since it was published in 1951. This is the famous Arrow Paradox. The essential ideas of this paradox of collective choice can be simply illustrated.[1]

As can be readily shown, sets of individual preferences may lead to unambiguous collective preference. In the situation shown in table 6-A-1, two thirds prefer A to C *and* two thirds prefer C to B. Thus, if the choice is between B and C and then between A and C, A will have a majority. But two thirds prefer B to A, and a different order of choice can produce any one of the choices. To the aficionados, this phenomenon is known as "cycling" or "intransitivity." It may be noted that the result is a consequence of the ordering of choices by one of the participants which leads to the preference for either extreme rather than a middle position. Thus, assume the order ABC proceeds along a spectrum politically. For concreteness, assume A is complete equality of income distribution, B is some redistribution, and C is no redistribution at all. In this case, individual 1 prefers complete equality but would rather have no redistribution at all than a moderate amount. The paradox has sometimes been cited as a

[1] See Arrow, *Social Choice,* Introduction to 2nd Edition. Arrow makes no claim that he was the discoverer of the paradox. Indeed, he traces the idea back to the Marquis de Condorcet in the eighteenth century. Arrow's rigorous methodology and attention to formal proof is, however, unique.

AUTHORS' NOTE: This appendix is based primarily on material prepared by Edwin T. Haefele.

TABLE 6-A-1. The Arrow Paradox

Individual	Individual's preference ordering				
1	A	→	C	→	B
2	C	→	B	→	A
3	B	→	A	→	C

demonstration that democratic choice cannot lead to an unambiguous selection of goals. The paradox, however, has been developed with mutually exclusive alternatives (either A or B or C but not A and B, and so forth), and democratic choice operates this way only in elections. Legislative bodies, which are the usual selectors of goals, normally have many issues under review at one time and do not form coalitions that persist over all issues (as game theorists normally assume), but rather form separate coalitions on different issues.[2]

The result of a formal proof that intransitivity may arise given certain conditions which Arrow regarded as reasonable is known as Arrow's Impossibility Theorem.[3] There are three very important conditions that he specified.

1. *Independence of Irrelevant Alternatives.* (pairwise comparisons) This condition says that the choice-making procedure should mean making choices between the alternatives x and y taken pairwise, and in the choice between x and y, the only circumstances taken into account shall be the relative positions of x and y on the preference schedules of the members of the group. In other words, the introduction of a third alternative z should not affect the relative preference for x and y. This condition has been critical in recent discussions and should be noted carefully by the reader.

2. *Monotonicity.* The essence of this condition can be stated as follows: if for a given group of preference schedules the social choice-making procedure means choosing x in preference to y, and if x is then displaced upward relative to y on some of the schedules in the original set, the procedure shall still choose x in preference to y.

3. *Unanimity.* If, in any set of schedules, x stands higher than y in the preference schedule of every member, the procedure shall mean choosing x rather than y.

[2] Different amounts of a publicly provided good or service are regarded as independent issues for present purposes.

[3] For a self-contained and relatively clear exposition of the theorem and its formal proof, see Duncan Black, "On Arrow's Impossibility Theorem," *The Journal of Law and Economics* vol. 12, no. 2 (October 1969).

Duncan Black, in the article on Arrow's Impossibility Theorem, has proven that if the condition of unanimity is imposed as a decision criterion (rather than, say, majority rule), intransitivity will arise very rarely, under the specified conditions, if the membership of the group is at all large. Of course, if mutually exclusive social states are being considered (and hence vote trading is ruled out by condition 1), very few people would presumably seriously propose the unanimity rule since it would virtually foreclose anything but negative decisions. It is reminiscent of the liberum veto which historians say wrought havoc in the Polish Diet from the mid-seventeenth to the end of the eighteenth century. Under this system, any motion to change the *status quo* was defeated if one member of the Diet shouted, "I disapprove!"

Haefele has analyzed the problem of collective choice in a more political context, first with mutually exclusive alternatives (elections) and then with independent issues in a legislative setting.[4] In brief, he has shown that, with independent issues, intransitivity does not arise given vote trading and majority rule. With mutually exclusive alternatives, he has shown that intransitivity does not occur under rules that approximate the two-party system. (The latter conclusion is consistent with Arrow's *possibility* theorem for two alternatives.) Finally, he has shown that, for given issues, both elections (following two-party rules) and legislatures (using vote trading), produce the same outcome. Thus he concludes that representative legislatures, using vote trading, can arrive at the same conclusions on issues as their constituents would have, had vote trading been possible for the latter, which it usually is not. The analysis implies that, even if it were possible, recourse to referendum-type decision making would be undesirable because it makes no provision for vote trading.[5]

The basic building block of the Haefele model is a vector of preferences related to a given set of independent issues, for example

$$\begin{bmatrix} Y_2 \\ N_1 \\ Y_3 \end{bmatrix}$$

which combines yes–no voting stances with an ordinal ranking of the importance of the issues to any individual. In the above case the individual

[4] Haefele, "A Utility Theory."

[5] "Madison, in Federalist Paper No. 55, expresses the point most succinctly, 'Had every Athenian citizen been a Socrates, every Athenian assembly would still have been a mob.' Need it be added that Madison's words relate to information costs, revealed preferences, and the lack of a vote-trading mechanism, or that modern proposals that everyone vote on all issues by electronic processes suffer the same defect as the Athenian assembly?" (Haefele, "A Utility Theory," p. 353).

is for the first issue, against the second, and for the third. The second is most important to him, the first next, and the third issue least important. A display of three such vectors (a three-man legislature) might give us

Issue	I	II	III		
A	Y_2	N_1	Y_1	$\rightarrow$	pass
B	N_1	Y_3	Y_3	$\rightarrow$	pass
C	Y_3	Y_2	Y_2	$\rightarrow$	pass

If votes are summed across rows, all three issues would be passed by the legislature (assuming majority rule) in this case. However, note that the first two men can trade votes on issues A and B.

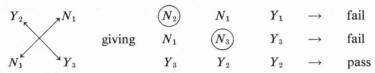

(N_2)	N_1	Y_1	$\rightarrow$	fail
N_1	(N_3)	Y_3	$\rightarrow$	fail
Y_3	Y_2	Y_2	$\rightarrow$	pass

making both better off. The outcome on the three issues now is that the first two fail and the third passes. The two men have improved their positions (both are better off) at the expense of the third, who is now worse off. (In larger matrices, some nontraders gain and others lose.)

Let us look again at the vectors:

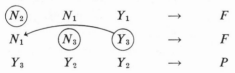

(N_2)	N_1	Y_1	$\rightarrow$	F
N_1	(N_3)	(Y_3)	$\rightarrow$	F
Y_3	Y_2	Y_2	$\rightarrow$	P

The third man will find it to his advantage to offer to change his vote on the middle issue if the first man will agree *not* to trade off his vote on the first issue. Then the vote will go

Y_2	N_1	Y_1	$\rightarrow$	P
N_1	Y_3	N_3	$\rightarrow$	F
Y_3	Y_2	Y_2	$\rightarrow$	P

If we take the three original vectors

Y_2	N_1	Y_1
N_1	Y_3	Y_3
Y_3	Y_2	Y_2

in the context of election strategy, we could imagine a candidate taking one of eight possible positions regarding the three issues:

Issue A	P	P	P	F	F	F	P	F
Issue B	P	P	F	P	F	P	F	F
Issue C	P	F	P	P	P	F	F	F

The implications of each position for each of the three voters is in the table below, which shows for each voter the importance to him of the issues that would pass for each of the possible candidates' positions.

Issue A	P	P	P	F	F	F	P	F
Issue B	P	P	F	P	F	P	F	F
Issue C	P	F	P	P	P	F	F	F
	Levels of Importance							
1st voter later wins[a] on issues of:	2nd 3rd	2nd	1st 2nd 3rd	3rd	1st 3rd	none	1st 2nd	1st
2nd voter later wins[a] on issues of:	2nd 3rd	3rd	2nd	1st 2nd 3rd	1st 2nd 3rd	1st 2nd 3rd	none	1st
3rd voter later wins[a] on issues of:	1st 2nd 3rd	1st 3rd	1st 2nd	2nd 3rd	2nd	3rd	1st	none

[a] For example, with candidates favoring the first position, $\begin{bmatrix} P \\ P \\ P \end{bmatrix}$, the first person would later win on issues that are of 2nd and 3rd degrees of importance to him—if the candidates of this persuasion won the election.

Opinion polls would reveal to the candidates that a majority favors passage of all three issues $\begin{bmatrix} P \\ P \\ P \end{bmatrix}$, yet inspection of the display above shows one vector $\begin{bmatrix} F \\ F \\ P \end{bmatrix}$ that can defeat $\begin{bmatrix} P \\ P \\ P \end{bmatrix}$ in majority voting Should one candidate choose $\begin{bmatrix} F \\ F \\ P \end{bmatrix}$, then the second candidate can defeat him by running on a platform $\begin{bmatrix} P \\ F \\ P \end{bmatrix}$, that is, the same outcome as the vote-trading outcome. Haefele demonstrates that this convergence holds for considerably more complicated cases, including the cyclical cases we referred to earlier.

What has been demonstrated here is that, in principle, a democratic legislature can make unambiguous choices which maximize the fulfillment of preferences, given a set of issues and vote trading.

CHAPTER 7

The Lower Delaware Valley Case Study

The residuals management and political modeling work done in the Quality of the Environment program, although productive of a number of results in its conceptual phases, led naturally to an attempt to apply it in a realistic setting. A detailed description of this effort was published in an RFF research paper *Environmental Quality Management: An Application to the Lower Delaware Valley,* by Walter O. Spofford, Jr., Clifford S. Russell, and Robert A. Kelly. This chapter has been adapted, and in many cases, quoted from that book.[1] The case study had several specific objectives. The order in which they are discussed does not imply relative importance.

One objective was to say something about the practical importance of including within a single analytical framework airborne, waterborne, and solid residuals. As we have shown in earlier chapters, the regional residuals management problem cannot, in principle, be solved by considering air or water or solids in isolation because of the links among forms of residuals and discharge media that are implied by the conservation of mass and energy in production, use, and residuals modification processes. There was, however, no hard evidence on the quantitative extent of the linkages and the size of the costs implied by isolated solutions in real situations. The applied model was intended to be a good enough representation of an actual region to yield one piece of defensible evidence, at least on the extent of the linkages.[2] This model is referred to here as the Lower Delaware Valley model, or the regional REQM model.

A second objective was to generate information on the cost implications—both total regional costs and distribution of costs—of utilizing

---

[1] Walter O. Spofford, Jr., Clifford S. Russell, and Robert A. Kelly, *Environmental Quality Management: An Application to the Lower Delaware Valley,* Research Paper R-1 (Washington, D.C., Resources for the Future, 1976).

[2] Available resources, including time, did not permit exploration of the second question.

various types of collective facilities within a regional REQM context. Examples of such facilities are regional sewage treatment plants, instream aeration installations, and plants to produce paper from used paper products generated in multiple activities.

A third objective was to explore the computational problems inherent in large-scale regional analyses, that is, how best to handle large amounts of data, how to aggregate with minimal loss of useful information, and how best to decompose large-scale models so that they become easier to compute.

A fourth objective was to produce an application with sufficient verisimilitude to see whether and how such models might work in an actual legislative and executive setting. Experimental work in this area was encouraging, but also indicated that the inclusion of the necessary information on the distribution of costs and benefits added significantly to the difficulties of constructing an applied regional REQM model and promised that in a real setting this information would add significantly to the model size. Making the extension include distributional information would permit an exploration of the degree to which the utility of the regional REQM model might be increased by the use of a political model that would allow potential legislative and executive users to explore distributional aspects efficiently. At the very least, this extension would permit the exploration of the costs of providing politically useful analysis on the distribution of costs and benefits of alternative physical measures for improving ambient environmental quality.

In the real world, the range of REQM strategies involved, the multiple sources and many types of residuals, and the varied distributions of effects and costs result in an almost infinite number of possible combinations. How can such complexity be represented and the results of analyses be communicated so as to make meaningful decisions possible? Developing data that would shed light on this question in a real situation was another closely related objective.

A final objective was to investigate the costs and benefits implied by incorporating a nonlinear model of the natural world, as discussed in chapter 5 and its appendixes, within a regional REQM model optimization framework. It was hoped to arrive at some basis for judging whether the increased computational costs and problems appear to be outweighed by some combination of greater accuracy and increased output of information relevant to policy making.

The Region

The region chosen for the application of the conceptual and mathematical model explained in chapter 5 and its appendixes was the Lower Delaware Valley. This is a very complex area with many individual point and nonpoint sources of residuals discharges. It was chosen in preference to simpler regions where fewer data might be required for several reasons. First, its very complexity provides a severe test of whether the analytical approach can be applied. If it can be applied there, it should be possible to apply it almost anywhere. In addition, the approach could represent the kernel of a useful analytical tool for policy making in one of the most important regions in the United States. Second, ties existed between RFF and certain regional institutions, particularly the Delaware River Basin Commission. Third, this region has been extensively studied for other purposes, so that a body of pertinent information was available surpassing that for most other regions. Accordingly, even though the Lower Delaware Valley is very complex, its advantages in terms of data availability and familiarity were considered sufficient to offset the costs of complexity.

The region is defined by county boundaries as shown in figure 20. It covers an area of about 4,700 square miles (approximately 12,200 square kilometers). The grid superimposed on the figure is used for locating dischargers and receptors of gaseous residuals in the model. It is related to the Universal Transverse Mercator Grid covering the United States. The region consists of Bucks, Montgomery, Chester, Delaware, and Philadelphia counties in Pennsylvania; Mercer, Burlington, Camden, Gloucester, and Salem counties in New Jersey; and New Castle County in Delaware. The major cities in the area are Philadelphia (coterminous with Philadelphia County); Trenton in Mercer County; Camden in Camden County; and Wilmington in New Castle County. There are 379 incorporated political jurisdictions—cities, towns, townships, boroughs, and divisions—in the eleven counties.

The 1970 population was a little more than 5.5 million. Of this, 35 percent was accounted for by Philadelphia alone, with Trenton, Camden, and Wilmington each accounting for 5 percent. However, these figures do not fully indicate the extent of urbanization, for ten of the eleven counties—Salem being the exception—are over 70 percent urbanized.[3]

[3] The definition of "urbanized" used by the U.S. Bureau of the Census is long and complex, but basically it amounts to counting as urbanized incorporated places of more than 2,500 inhabitants or of a density greater than 1,000 persons per square mile.

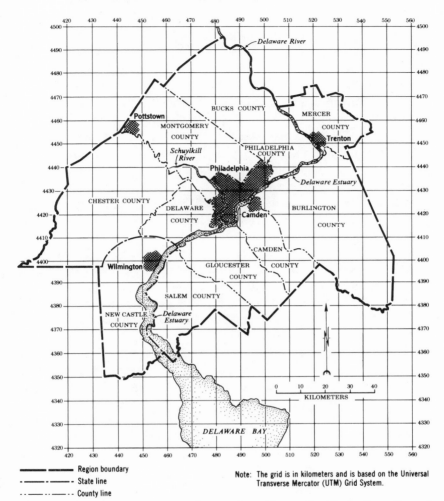

Figure 20. Lower Delaware Valley region. From: Spofford, Russell, and Kelly, *Environmental Quality,* p. 4.

Population densities range to as high as over 15,000 per square mile in Philadelphia, and are 1,000 or greater for four other counties. In both Pennsylvania and Delaware, densities in the region are uniformly greater than the average for the respective states as a whole, but in New Jersey the suburban counties of Burlington and Gloucester, and rural Salem County are all less densely populated than the average for the state.

Incomes are generally high in the region. Using median family income as the indicator, every county except Philadelphia has a higher median family income than the United States as a whole. Salem County has the lowest median income outside Philadelphia; Montgomery County has the region's highest median income. The difference between these two extremes is about $2,500 per family per year, or about 25 percent of the lower figure. Thus, there are significant intraregional income differences, even at the highly aggregated level of counties. This suggests that the distributional implications of REQM strategies will be an important issue in the region. This presumption is accentuated by the fact that it is Philadelphia that sits in the center of the largest industrial concentration, has the highest population density, and hence can be expected to have the greatest ambient environmental quality (AEQ) problems, that also has the lowest per capita income base from which to pay for improving AEQ.

The region is one of the most heavily industrialized areas in the United States. It contains: seven large petroleum refineries; five large steel mills and many smaller ones; thirteen pulp and paper or paper mills with more than 100 tons per day output, and numerous smaller ones; fifteen large thermal electric energy-generating plants and two smaller ones; many large and small chemical and petrochemical plants of various types; and many other types of industrial operations, including foundries and automobile and electronic assembly plants.

In 1968, there were seventeen municipal sewage treatment plants in the region with flows greater than one million gallons per day (mgd) discharging to the Delaware Estuary. (Seven of these plants had flows greater than 10 mgd.) In addition, 123 sewage treatment plants of various sizes discharged to the Schuylkill River and to the tributaries of the two major rivers.[4] In 1970 there were seventeen municipal incinerators and numerous landfills and dumps in operation in the region.

Surface Waters

The major recipient of waterborne residuals in the Lower Delaware Valley region is the Delaware Estuary itself.[5] The estuary consists of the 85-mile stretch between the head of the tide at Trenton and the head of

---

[4] In the modeling application, 1970 was used as the base period, and the model structure reflects the situation in that year whenever possible. In some cases, such as sewage treatment plants, in which government surveys were the data source, the base year differs from 1970.

[5] There are significant groundwater resources in the basin, particularly in the coastal plain section of Delaware. The Delaware River in the reach from Trenton to Philadelphia is the major source of recharge for the Magothy and Raritan aquifers, which provide significant sources of water for users in New Jersey. These and other aspects of groundwater in the region were not explicitly considered in the model.

Figure 21. Reach locations for the Delaware Estuary model. From: Robert A. Kelly, "Conceptual Ecological Model of the Delaware Estuary," in B. C. Patten, ed., *Systems Analysis and Simulation in Ecology,* vol. IV (New York, Academic Press, 1976) (reprinted by permission of Academic Press and the author).

Delaware Bay at Liston Point, Delaware. Figure 21 shows the estuary and twenty of the twenty-two reaches into which it was divided for modeling purposes.

The flow of the river varies widely, from month to month and year to year. During the period from 1951 through 1972, for example, the monthly mean flow at Trenton varied from a low of about 1,550 cubic feet per second (cfs) in July of 1965 to a high of nearly 40,000 cfs in late spring of 1958. The mean annual flow for this period ranged from a low of about 4,700 cfs for 1965 to a high of about 18,000 cfs for 1952. The mean discharge at Trenton over the sixty-year period from 1912 to 1972 was about 11,400 cfs.[6] The low flow period, and hence the period during which the river is most sensitive to residuals discharges, is July through

[6] Information provided by the Water Resources Division, U.S. Geological Survey, Reston, Virginia.

TABLE 29. Estimated Residuals Discharges to the Delaware Estuary, September 1970

Sources	Biochemical oxygen demand		Nitrogen		Phosphorous	
	Lbs/day	Percentage of total	Lbs/day	Percentage of total	Lbs/day	Percentage of total
Industrial	385	41	117	53	7.93	15
Municipal	425	46	63.8	29	27.0	50
Tributaries	50	5	19.1	9	13.0	24
Storm water	76	8	20.5	9	5.76	11
Total (rounded)	936		220		53.7	

Distribution of September 1970 residuals discharges
83% of BOD_5 load is discharged into 27% of reaches
79% of N load is discharged into 27% of reaches
80% of P load is discharged into 27% of reaches

Measured ambient water quality resulting from September 1970 residuals discharges (mg/l)
Minimum mean dissolved oxygen concentration: 1.2
Maximum mean BOD_5 concentration: 6.4
Maximum mean P (as P) concentration: 0.33
Maximum mean N (as TKN)[a] concentration: 2.5

Source: Blair T. Bower, ed., *Regional Residuals–Environmental Quality Management Modeling*, Research Paper R-7 (Washington, D.C., Resources for the Future, 1977) p. 61.
  [a] NH_3 + organic N.

October. The flow of the Delaware River at Trenton, used as the basis for the aquatic ecosystem model described later, corresponds to the mean flow for September 1970, that is, about 4,100 cfs. This flow is about 2.8 times as large as the seven-day, ten-year flow at Trenton of 1,500 cfs. Table 29 shows the estimated residuals discharges to the twenty-two reaches of the estuary for September 1970, and the heavy concentration of those discharges in a relatively few reaches. The maximum and minimum concentrations in the estuary for that month are also shown.

The Schuylkill River is considerably smaller than the Delaware River. The mean discharge for the forty-seven-year period from 1926 to 1972 was about 1,800 cfs at Pottstown, Pennsylvania, approximately where the Schuylkill enters the Lower Delaware Valley region. (The mean discharge for the Schuylkill at Philadelphia over a thirty-four period was approximately 2,850 cfs.) The mean annual flow at Pottstown for the period 1951 through 1965 varied from a low of 840 cfs for 1965 to a high of 3,060 cfs for 1952. The monthly mean flow during this same period ranged from a low of 320 cfs in October 1963 to a high of 4,760 in February

1951. The seven-day, ten-year flow of the Schuylkill at Pottstown is about 240 cfs.

The major tributaries in the Schuylkill-Delaware river basins are still smaller, their mean annual flows ranging from a high of about 430 cfs for Brandywine Creek in Wilmington, Delaware, to a low of about 4 cfs for Blackbird Creek in Blackbird, Delaware. The flows in many of these tributaries are so low in some months of the summer that return flows from wastewater treatment plants currently comprise a major portion of the flows in the tributaries during low flow periods.

Atmosphere

The atmospheric "resource" of the region does not lend itself to as simple a characterization as the flow relations for its watercourses. For the region as a whole, the seasonal prevailing wind pattern is roughly: winter and spring westerlies (and west-northwesterlies); summer southwesterlies; and autumn variability. In almost every month there are southwesterly winds along the stretch from roughly the Delaware state line to Camden, New Jersey. The net effect of these meteorological conditions, together with the spatial pattern of discharges of gaseous residuals, is a "mountain" of poor quality air along the estuary from Wilmington to Trenton, with the highest concentrations over the Philadelphia-Camden area.

During the period from 1936 to 1965, atmospheric stagnations[7] over Philadelphia occurred most often during the month of September, with sixteen cases recorded out of a total of thirty-three; the next most frequent month was October with eight cases.[8] Episodes of poor air quality are often associated with atmospheric stagnations, although they are also dependent on residuals discharges during these adverse meteorological periods. During the two-year period 1957–1959, twenty-two "air pollution episodes" were recorded in the Philadelphia area—four in the spring, two in the summer, eight in the fall, and eight in the winter.[9] From information on the frequency of atmospheric stagnations alone, proportion-

[7] Atmospheric stagnations are defined here as periods of four or more days when the windspeed at anemometer level does not exceed 7.5 knots per hour.

[8] J. Korshover, *Climatology of Stagnating Anticyclones East of the Rocky Mountains, 1936–1965,* U.S. Department of Health, Education and Welfare, Public Health Service, Publication No. 999-AP-34 (1967).

[9] F. K. Davis, "The Air Over Philadelphia," *Symposium: Air Over Cities,* U.S. Department of Health, Education and Welfare, Public Health Service, Technical Report A62-5 (November 1961) pp. 115–129.

TABLE 30.   Estimated Sulfur Dioxide and Particulate Discharges in the
Lower Delaware Valley Region, 1970
(annual average in tons per day)

Discharger category	Sulfur dioxide	Particulates
Point sources (1,031 stacks)		
Petroleum refineries	410	66
Steel mills	19	58
Power plants	1,332	126
Other point sources	439	133
Total	2,200	383
Area sources (240 areas)		
Household heating[a]	214	25
Other area sources	550	191
Total	764	216
Total, all sources	2,964	599

Source: Based on EPA's 1970 inventory of gaseous discharges for the Metropolitan Philadelphia Interstate Air Quality Control Region, supplied by the Division of Applied Technology, Office of Air Programs, U.S. Environmental Protection Agency, Durham, North Carolina, and appearing in Spofford, Russell, and Kelly, *Environmental Quality*, p. 15.

[a] Estimated from the number of housing units and the household heating fuel types contained in the 1970 Bureau of the Census computer tapes.

ally more episodes would have been expected in the fall and fewer in the winter. However, discharges of sulfur dioxide and particulates in the the Philadelphia area are typically higher in the winter than during any other period of the year.

During the 1967–1968 period, the maximum annual average concentrations of sulfur dioxide (SO_2) and total suspended particulates (TSP) in the region were about 190 micrograms per cubic meter ($\mu g/m^3$) and about $150 \mu g/m^3$, respectively, measured at a station in Philadelphia. These compare with the U.S. primary ambient standards of $80 \mu g/m^3$ and $75 \mu g/m^3$, respectively. Table 30 shows the estimated discharges of SO_2 and particulates in the Lower Delaware Valley region for 1970.

The major discharges of SO_2 and particulates in the region are the petroleum refineries, steel mills, and power plants. Collectively, their SO_2 discharges in 1970 averaged about 1,760 tons per day, roughly 60 percent of the total of about 3,000 tons per day. Their collective discharges of particulates averaged about 250 tons per day, roughly 40 percent of the total of about 600 tons per day for the region. The power plants in the region clearly represent the largest set of dischargers of both SO_2 and particulates.

In 1970, area sources of SO_2 and particulate discharges accounted for about 25 percent and 35 percent respectively of the totals for the region. However, these sources typically discharge close to the ground and hence contribute proportionally more to ground level ambient concentrations than the relative magnitudes of their discharges would indicate. "Area sources" of gaseous residuals are not tied to specific stack locations but were analyzed as though their discharges were uniformly generated over the different subareas identified.

For modeling air quality, the atmospheric conditions used represent the annual joint probability distribution of wind speed, wind direction, and stability conditions for 1970, assumed to be uniform throughout the region.[10] For neither air nor water quality analyses, were conditions representing rare events used in the analyses. Ideally, explicit attention would also have been given to this aspect of regional REQM.

The Regional REQM Model

The framework of the regional REQM model of the Delaware Valley Estuary is shown in figure 22. There are three main parts of this model: a linear programming model of residuals generation and discharge (comprising both production and use activities), the natural systems models, and the environmental evaluation section. The regional REQM model is designed to provide at the least cost possible a way of: (1) producing an exogenously determined "bill of goods" at the individual industrial plants; (2) meeting electrical energy and household and commercial space heating requirements for the region; and (3) handling, modifying, and disposing of specified quantities of municipal liquid and solid residuals, subject to constraints on (a) the distribution of ambient environmental quality—water, air, and landfills—over geographic units; and (b) the distribution of consumer costs—electrical energy, heating fuel, sewage disposal, solid waste disposal, and regional instream aeration—over political jurisdictions.

Constraints (1), (2), and (3) can be exemplified by maximum concentrations of SO_2 and TSP at a number of receptor locations in the region, minimum concentrations of dissolved oxygen and fish biomass in the estuary, maximum concentration of algae in the estuary, and restrictions on the types of landfill operations that can be used in the region. With respect to the latter category of constraints, the model permits constraining increases in: the cost of electricity by utility service area caused

───────────

[10] This use of uniform wind rose and stability conditions for the region, in combination with the available source inventory data, produced distributions of ambient air quality similar to the measured annual averages.

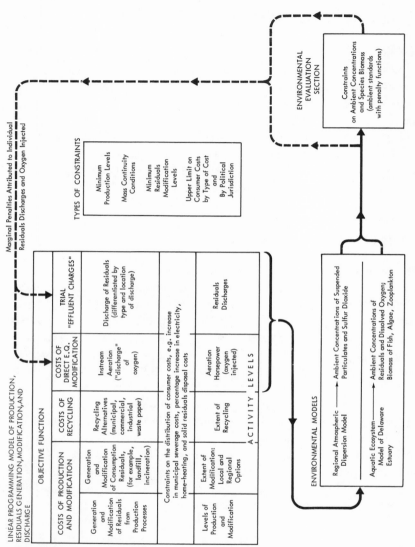

Figure 22. A regional REQM model. From: Spofford, Russell, and Kelly, *Environmental Quality,* p. 22.

by all required discharge reduction activities, the cost of household heating caused by fuel switching, and municipal expenditures caused by increased liquid residuals modification and more expensive solid residuals management methods. However, credit is given for the sale of newsprint and linerboard produced from used newspapers and used corrugated containers, respectively, collected within the region, such production being an alternative to solid residuals disposal to landfills, incineration, or both.

Referring to figure 22, in the upper left-hand block of the diagram is the basic driving force for the entire model, a linear programming model of residuals generation and discharge. It is in this part of the model that the minimum "production" constraints are found. A key output of this part is a vector of residuals discharges identified by substance and location. This vector is fed into the environmental models, the model of the aquatic ecosystem, and the dispersion model for SO_2 and TSP. This section of the overall model, in turn, produces as output a vector of ambient environmental quality levels (for example, SO_2 in the thirty-seventh political jurisdiction). These concentrations are then treated as input to the "evaluation" submodel found in the lower right of the figure. Here the concentrations implied by one solution of the residuals generation and discharge submodel are compared with the exogenously established ambient environmental standards imposed for the model run. An iterative procedure, based on the gradient method of nonlinear programming and on the use of penalty functions, was developed to meet the ambient standards within some specified tolerance, and to select a vector of residuals discharges that meets the constraints at least cost to the region.

The following are the main features of the Lower Delaware Valley model.

1. It is a mathematical model. It is intended to provide least cost ways of achieving various levels of AEQ subject to constraints on production and use, AEQ, and consumer costs.
2. It reflects some of the nontreatment alternatives available for reducing the amounts of residuals initially generated, especially in production processes. In general, the amount (or form) of residual to be handled is not assumed to be fixed.
3. It deals with the three major forms of materials (liquids, gases, and solids), heat, and the three receiving environmental media (watercourses, atmosphere, and land) simultaneously. The model reflects the conservation of mass and energy for the relevant residuals, accounting for material and energy flows as they are modified from one form to another in production, use, and residuals modification activities. Carbon dioxide and water vapor are ignored.
4. It is capable of incorporating various types of models of the natural world from the simplest linear transformations—for ex-

ample, steady state Streeter-Phelps-type dissolved oxygen models
—to complex simulation models of aquatic ecosystems.

5. It is a static, economic model so that time is not considered in the residuals generation and discharge portion of the model; capacity expansion and optimal timing are not explored; no new industries come into the region; and the population level and its distribution remain constant. The model output represents a "snap shot" of the region for fixed conditions at a given point in time. A dynamic analysis of the effect on the region of economic and population growth cannot be made, but a comparative static analysis can be made.

6. The spatial distribution of activities in the region is fixed, although the effects of various distributions can be examined and compared.

7. Production and use activities in the region are deterministic and steady state.

8. The natural systems models are deterministic and steady state, although the ecosystem model is structured as a nonsteady state model, but the eventual steady state results are used in the calculations.

9. A single season (spring, summer, fall, winter, annual) is employed for the analysis, although the model can be operated for different seasons or time periods and the optimal solutions for each season compared.

10. Interactions among residuals in the atmosphere are assumed not to occur, and decay rates, where applicable, are independent of the quantities of the residual present and of the presence of other residuals.

To provide specific information on the geographic distribution of ambient environmental quality and of REQM costs to consumers throughout the region, the Lower Delaware Valley was divided into fifty-seven political jurisdictions with 1970 populations ranging from approximately 60,000 to 110,000. The objective was to make each jurisdiction as socially and economically homogenous as possible. To form these jurisdictions, some of the 379 cities, towns, boroughs, and townships that comprise the region were aggregated and others were subdivided; a few areas, such as major parks and military bases, were excluded. However, all fifty-seven political jurisdictions are comprised of whole census tracts and are located entirely within the boundaries of

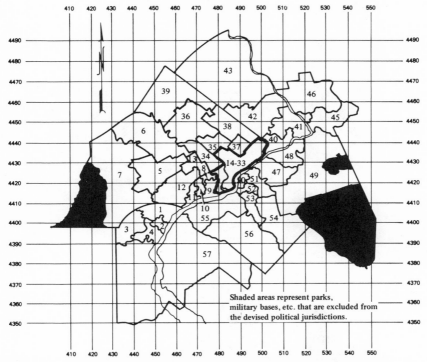

Figure 23. Political jurisdictions devised for the Lower Delaware Valley region

individual counties and hence within states. Figure 23 shows the jurisdictions outside Philadelphia; figure 24 those within Philadelphia.

For each political jurisdiction the following measures of ambient environmental quality and of increased consumer costs resulting from REQM are available: air quality (annual average ground level concentrations of SO_2 and TSP at an arbitrarily selected location within each jurisdiction); landfill quality; and levels of increased consumer costs (uniform over each jurisdiction) for electrical energy, household and commercial heating, municipal sewage handling (including modification and disposal), and municipal solid residuals disposal. Ambient water quality in the estuary—in terms of concentration of dissolved oxygen, biomass of fish, and biomass of algae—is available for each of the twenty-two reaches shown in figure 21 and is related to the fifty-seven political jurisdictions only indirectly. The allowable upper (or lower if appropriate) limits of these AEQ indicators can be constrained.

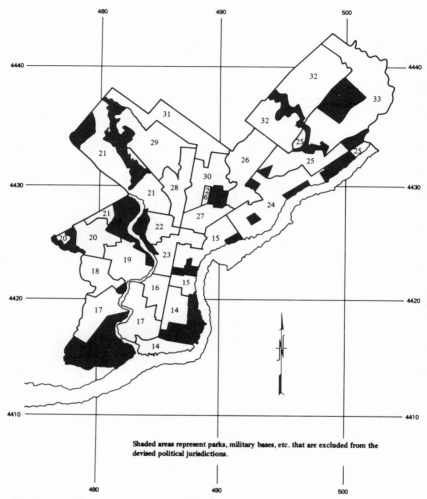

Shaded areas represent parks, military bases, etc. that are excluded from the
devised political jurisdictions.

Figure 24. Political jurisdictions devised for the Philadelphia portion of the
Lower Delaware Valley

The estuary is the only water body in the region that is described by
a water quality model. Dischargers in the reach of the Schuylkill River
included in the model are handled with specified discharge standards
based approximately on state standards existing in 1968–70.[11] This

[11] Resources were not available to develop a water quality model of the Schuylkill River. Development of such a model was subsequently undertaken for the Environmental Protection Agency but was not completed in time to be utilized in the
RFF work.

arrangement does not permit liquid discharge tradeoffs in this part of the region. Dischargers to other tributaries of the Delaware Estuary are dealt with in the same manner as dischargers to the Schuylkill. The stretch of the Delaware River above Trenton currently is of fairly good quality, and there appears to be little likelihood of major deterioration in the near future. Consequently, no water quality model of this part of the region was considered necessary.

Residuals Generation and Discharge Models

The residuals generation, modification, and discharge submodel of the Lower Delaware Valley model is composed of six separate linear programming models arranged in individual modules. These modules are depicted in table 31. The left-hand side of the table indicates the module number, for identification purposes only. (The MPSX designation derives from the particular computational routine used in the analysis.) Columns 1, 2, and 3 indicate the sizes of the individual linear programming modules—number of rows, number of columns, and number of residuals discharges, respectively. In total, in the model there are: almost 8,000 variables (columns); a little over 3,000 constraining relationships (rows); and almost 800 individual residuals discharges that enter the natural systems models. The residuals involved are biochemical oxygen demand (BOD_5), nitrogen, phosphorus, toxics (for which phenols comprise the proxy), suspended solids, and heat discharges to the estuary, and SO_2 and particulate discharges to the atmosphere. The fourth column describes the type and indicates the number (in parentheses) of activities in the region for which residuals management options are included in the model. The fifth column lists the information generated by the model on the distribution of extra costs for REQM. Except for the costs of sewage disposal, there is one constraint for each type of extra cost and for each of the fifty-seven political jurisdictions into which the Lower Delaware Valley has been divided. For sewage disposal, there are only forty-six extra costs in the model, because eleven jurisdictions do not discharge any sewage directly to the estuary.

In the model, 297 point and nonpoint sources of gaseous and liquid residuals are provided with management options for reducing their discharges. The 183 point sources include 124 industrial plants and 59 municipal residuals-handling and disposal activities—23 incinerators and 36 sewage treatment plants. The 124 industrial plants in turn include, among others, the 7 oil refineries in the region, 5 steel plants, 17 power

TABLE 31.   Lower Delaware Valley Model: Residuals Generation and Discharge Modules

Module identifi- cation	Size of linear program			Description (4)	Percent extra costs for the 57 political jurisdictions (except as noted) (5)
	Rows (1)	Col- umns (2)	Dis- charges (3)		
MPSX 1	286	1,649	130	Petroleum refineries (7) Steel mills (5) Powerplants (17)	57 electricity
MPSX 2	741	1,482	114	Household heat (57) Commercial heat (57)	57 fuel 57 fuel[a]
MPSX 3	564	1,854	157	"Over 25 $\mu g/m^3$" dischargers (75)[b]	
MPSX 4	468	570	180	Delaware Estuary sewage treatment plants (36)	46 sewage disposal ($ per house- hold per year)[c]
MPSX 5	951	1,914	88	Paper plants (10) Municipal incinerators (23) Municipal solid residuals handling and disposal activities	57 solid residuals disposal
MPSX 6	229	395	117[d]	Delaware Estuary industrial dischargers (23)[e] Instream aeration (22)	57 instream aera- tion (absolute extra cost per day)[f]
Total	3,239	7,864	786		

Note: MPSX is a designation that derives from the particular computational routine used in the analysis. Figures in parentheses are the number of activities in the region for which residuals management options are included in the model.

Source: Spofford, Russell, and Kelly, *Environmental Quality* p. 29. The major sources of information and data used in the construction of the residuals generation and discharge models were: (1) EPA's 1970 inventory of SO_2 and particulate discharges in the Metropolitan Philadelphia Interstate Air Quality Control Region; (2) Delaware River Basin Commission for the municipal and industrial wastewater dischargers to the Delaware Estuary; (3) Regional Conference of Elected Officials' publications for municipal incinerators and for electricity districts and power plants; and (4) the 1970 U.S. Bureau of Census computer tapes for number of people and housing units by political jurisdictions, household heating fuel characteristics, and sewerage systems. The base period for data employed in this model is 1968 to 1970.

  [a] The commercial heating requirements in this module are based on the differences between SO_2 discharges from area sources in the IPP inventory of gaseous discharges and SO_2 discharges from the household heating model. Commercial heating requirements for political jurisdictions 19, 28, 29, 31, 32, 33, 37, and 44 are equal to zero.

  [b] Includes each industrial plant whose gaseous residuals discharges result in maximum annual average ground level concentrations equal to or greater than 25 $\mu g/m^3$. To determine this group, all stacks in the Implementation Planning Program (IPP) inventory were considered. The maximum annual average ground level concentrations of SO_2 and TSP were computed for each stack. For all stacks at the same x-y location (i.e., same plant), the maximum ground level concentrations were added together. Those plants resulting in maximum ground level concentrations, for either SO_2 or TSP, equal to or greater than 25 $\mu g/m^3$ were then placed in this category.

  [c] The extra costs reported represent the *average* increase per household per year for each political jurisdiction. Jurisdictions 5, 6, 7, 36, 39, 43, 45, 46, 53, 54, and 56 do not discharge to the estuary at all.

plants, and 23 other industrial dischargers to the estuary. The 114 area (nonpoint) sources are 57 household heating activities and 57 commercial heating activities (one of each for each jurisdiction). The other point and nonpoint sources identified in the region that are not provided with residuals management options are incorporated as background dischargers.

The management options available to the various sources for reducing their discharges range from alternative production processes and raw material inputs to residuals modification at the "end of the pipe." Examples of these options are shown in table 32. In addition to the management options for the individual activities listed in the previous paragraph, four collective residuals-handling and modification activities, and one direct modification of AEQ activity are included: incineration, landfilling, regional sewage modification, and recycling of used newspapers and used corrugated containers represent the former; instream aeration, the latter.

The sources of residuals discharges included in the regional model do not encompass all residuals dischargers in the Lower Delaware Valley. Certain kinds of activities and the residuals generated by and discharged from them have not been included (for example, transportation, agricultural activities, and construction activities), except as background sources of residuals where appropriate. For certain kinds of AEQ problems, these sources are important, and they could have been analyzed in an analogous manner. However, the activities that have been included in the model include the important dischargers of SO_2 and particulates to the atmosphere and the major dischargers of BOD_5, nitrogen, and phosphorus to the Delaware Estuary.

Natural Systems Models

The second major component of the regional REQM model consists of the natural systems models of the Delaware Estuary and of the air over the region. The former is a nonlinear ecosystem model of the Delaware Estuary, divided into twenty-two reaches as shown in figure 21. (A general description of the model is in chapter 5, with a more detailed descrip-

---

d Does not include the 22 oxygen "discharges" from the instream aeration option.

e Twelve of the Delaware Estuary industrial wastewater dischargers in MPSX 6 are also represented by SO_2 and/or TSP dischargers in MPSX 3, and the gaseous residuals discharges of another are included in MPSX 5.

f The model currently reports the total regional absolute extra cost per day for instream aeration. The cost is then allocated equally among the 57 political jurisdictions. Any other distribution is also possible.

TABLE 32. Residuals Management Options Available to Selected Types of Dischargers in the Lower Delaware Valley Model

Type of residual discharger	Management option available	Primary residual reduced	Secondary residual generated[a]
Petroleum refineries	1. Secondary or tertiary wastewater treatment, and various reuse alternatives (cooling tower water makeup, desalter water, boiler feedwater)	BOD$_5$	Sludge
	2. Cooling tower	Heat	[b]
	3. Burn lower sulfur fuel	SO$_2$	None
	4. Refine lower sulfur crude	SO$_2$	None
	5. Cyclone collectors on cat-cracker catalyst regenerator (2 efficiencies: 70%, 85%)	Particulates	Fly ash
	6. Electrostatic precipitator, 95% efficiency	Particulates	Fly ash
	7. Sell, rather than burn, high sulfur refinery coke	SO$_2$ and particulates	None
	8. Sludge digestion and landfill	Sludge	[c]
Thermal power generating facilities	1. Cooling tower	Heat	[b]
	2. Burn lower sulfur coal	SO$_2$	None
	3. Limestone injection-wet scrubber, 90% efficiency	SO$_2$	Slurry
	4. Electrostatic precipitators (3 efficiencies: 90%, 95%, 98%)	Particulates	Fly ash
	5. Settling pond, 90% efficiency	Slurry	Solid "ash"
Municipal sewage treatment plants	1. Secondary or tertiary wastewater treatment	BOD$_5$	Sludge
	2. Sludge digestion, drying, and landfill	Sludge	[c]
	3. Sludge dewatering and incineration	Sludge	Particulates, Fly ash
	4. Dry cyclone collector (2 efficiencies: 80%, 95%)	Particulates	Fly ash
Municipal incinerators	1. Electrostatic precipitators (2 efficiencies: 80%, 95%)	Particulates	Fly ash

Source: Bower, *Regional Residuals* p. 73.
[a] A secondary residual is generated in the process of modifying a primary residual.
[b] Heat is rejected to the atmosphere along with water vapor.
[c] The secondary residual is digested sludge at a different location.

tion in appendix 5-B.) Inputs of liquid residuals to the estuary model are: organics as biochemical oxygen demand (BOD_5), nitrogen, phosphorus, toxics (for which phenols are the proxy), suspended solids, and heat (in Btu). Outputs are expressed in terms of ambient concentrations of algae, bacteria, zooplankton, resident fish, dissolved oxygen, BOD_5, nitrogen, phosphorus, toxics, suspended solids, and temperature. Three of these outputs—algae, fish, and oxygen—can be constrained. These constraint levels are set exogenously to the regional REQM model and represent both a major driving force in the solution runs and one of the principal issues the REQM model is designed to study. Thus, the costs of meeting alternative sets of AEQ levels can be viewed as the most important single output capability of the regional model. A diagram showing the structure of the Delaware Estuary ecosystem model in terms of materials flows is shown in figure 5-B-2, and a summary of the model indicating relevant inputs and outputs is presented in table 33.

With respect to air quality, the regional Delaware REQM model includes two 57 × 240 (57 receptor locations, one for each political jurisdiction, and 240 dischargers) air dispersion matrices—one each for SO_2 and TSP. These matrices relate annual average ambient ground level concentrations to discharges of SO_2 and particulates. The matrices are based on output of the air dispersion model contained in EPA's Implementation Planning Program (IPP). The air dispersion model was calibrated using equations developed in an unpublished EPA study of the Philadelphia Metropolitan Air Quality Control Region. A summary of the air dispersion model, indicating relevant inputs and outputs, is shown in table 34.

To keep the two air dispersion matrices as small and manageable as possible, all stacks at the same plant location in EPA's inventory of gaseous residuals discharges were aggregated to a single stack. This required computing the characteristics of a single stack that produced the same result as the group of stacks, that is, the same maximum ground level concentration as the group, using discharge rates from EPA's inventory. The discharge rate employed in the analysis for the single stack was the aggregate discharge of all the individual stacks.

Stack characteristics of the single stack (physical stack height, plume rise, and effective stack height) were determined for discharges of both SO_2 and particulates. In general, the two "artificial" stacks computed on the basis of the above criterion differed substantially because the characteristics of the equivalent single stack depend on the relative discharge

TABLE 33. Delaware Estuary Ecosystem Model

Endogenous variables (compartments—mg /l)
  Algae
  Zooplankton (herbivores, detritivores, and bacterivores)
  Bacteria
  Fish
  Dissolved oxygen (DO)
  Organic matter (as BOD_5)
  Nitrogen
  Phosphorus
  Toxics
  Suspended solids
  Temperature (°C)

Inputs of Residuals (lbs /day)	*Target Outputs*[a] (mg /l)
Organic material (as BOD_5)	Algae
Nitrogen	Fish
Phosphorus	Dissolved oxygen
Toxics	
Suspended solids	
Heat (Btu)	

Model
  Type: materials balance-trophic level
  Characteristics: deterministic, nonsteady state
  Calibration: based on average September 1970 flow at
    Trenton, N.J., of 4,150 cfs.
Reaches
  Number: 22

Source: Based on Kelly, "Conceptual Ecological Model"; and Robert A. Kelly and Walter O. Spofford, Jr., "Application of an Ecosystem Model to Water Quality Management: The Delaware Estuary," in C. A. S. Hall and J. W. Day, Jr., eds., *Ecosystem Modeling in Theory and Practice: An Introduction with Case Histories* (New York, Wiley, 1977).

[a] The regional REQM model is operated for relevant minimum, or maximum, allowable ambient concentrations ("standards").

rates among the individual stacks. In developing the two air dispersion matrices, two different sets of single stacks were used, one for the SO_2 analysis and another for the particulates analysis.

Setting Up the Model Runs

The operational objective in running the regional REQM model was to explore the effects on aggregate regional REQM costs and the distribution of those costs of the following decision variables: alternative requirements in the quality of landfills; alternative ambient water and air quality standards; an instream aeration option for directly improving estuary water quality; a regional sewage treatment plant option; and a regional

TABLE 34.   Air Dispersion Model

Endogenous variables (ground level, annual average ambient concentrations—μg/m³)
  Sulfur dioxide
  Suspended particulates

Inputs of residuals (tons/day)	*Target outputs*[a] (μg/m³)
Sulfur dioxide	Sulfur dioxide
Particulate matter	Total suspended particulates

Model

  Type: Gaussian plume dispersion model from EPA's Air Quality Implementation
  Planning Program (IPP)

  Characteristics: deterministic, steady state

  Input requirements:

1. *Sources*	3. *Meteorological conditions*
For each stack:	a. Annual joint probability distribution
a. *x-y* coordinates	for wind speed (6 classes), wind direc-
b. Physical stack height	tion (16 directions), and atmospheric
c. Stack diameter	stability (5 classes), using 1970 data
d. Discharge rate for each	for Philadelphia; the result is 480
gaseous residual stream	discrete meteorological situations
with associated exit tem-	b. Mean temperature: 68°F (20°C)[b]
perature and exit velocity	c. Mean annual pressure: 1,017 millibars
2. *Receptors*	(30.3 inches of mercury)
x-y coordinates of each receptor	d. Mean maximum afternoon atmos-
location	pheric mixing depth: 1,000 meters[c]

Source: TRW (Thompson Ramo Woolridge), Inc., *Air Quality Implementation Planning Program*, vols. I and II (Washington, D.C., U.S. Environmental Protection Agency, November 1970) (also available from National Technical Information Service, Springfield, VA 22161, PB-198 299 and PB-198 300).

[a] The regional REQM model is operated for maximum allowable ambient concentrations ("standards").

[b] The mean temperature used in the model, 68°F, is roughly that for September. The mean annual temperature is 64°F.

[c] The mean maximum afternoon mixing depth used in the model, 1,000 meters, is the average for the months of May, August, and September. In June and July, the mean mixing depth is greater than 1,000 meters; from October to April, it is less than 1,000 meters. The effect of decreasing the mixing depth around the 1,000 meter level is to increase the concentrations predicted by the model.

recycling option for solid residuals management. For both air and water quality, there are two alternative sets of standards, a relatively easy (E) set to meet and a tight (T) set. For landfill operations, three alternative quality levels are specified: low, medium, and high. The sets of standards for the three media that are used in various combinations in the analyses, are shown in table 35.

Analysis with the model involves specifying for each production run a combination of ambient environmental quality standards to be met and permitted regional residuals-handling and modification options. Most runs were arbitrarily stopped at between 29 and 35 iterations, although technically the optimum, even a local optimum, was not necessarily

TABLE 35. Air, Water, and Land Environmental Quality Standards for Production Runs of the Lower Delaware Valley Model

Qualities	Easy standards (E)	Tight standards (T)
Water quality (milligrams per liter)		
Algae	3.00	2.00
Fish	0.01	0.03
Dissolved oxygen	3.00	5.00
Air quality (micrograms per cubic meter)		
Sulfur dioxide	120	80[a]
Suspended particulates	120	75[a]
Landfill quality		
L (low)	Open dump, but no burning allowed.	
M (medium)	Good quality sanitary landfill.	
H (high)	Good quality sanitary landfill with shredding, impervious layer to protect groundwater; treatment of leachate; measures to enhance aesthetics, such as fences, trees, etc.	

Source: Spofford, Russell, and Kelly, *Environmental Quality*, p. 52.

[a] For certain jurisdictions, it is technically impossible, as the model is constructed, to meet the stricter standard. Consequently, the following modifications were made to the stricter standards:

Political jurisdiction	TSP ($\mu g/m^3$)	Political jurisdiction	SO₂ ($\mu g/m^3$)
14	76	16	83
15	82		
22	76		
23	79		
24	79		
27	81		
50	76		

achieved.[12] However, experience indicated that at this number of iterations the ambient environmental constraints are met, if it is technically possible to meet them. This location is high up on the response surface and continuing beyond this point improves the objective function value only slightly. This is not meant to imply, however, that residuals discharge *patterns* would not change, perhaps even substantially if the run were to be continued beyond 30 iterations. Experiments in running the model beyond 30 iterations indicated that, although the *total* residuals discharges do not change much beyond the 30 to 35 iterations (because the

[12] At 1975 computer rates, it cost $1,220 for a 30-iteration run, or roughly $41 per iteration, operating on an IBM 370 model 165 and using 300K bytes of internal core storage. On average, one iteration takes 2.80 minutes of computer-processing unit time, 5.94 minutes of input–output time, has 13,400 read-write instructions, and prints 1,236 lines of output (minimum).

ambient standards are being met), tradeoffs in reductions among individual dischargers are still taking place.[13] Thus, a very relevant question is: how comparable are the results—costs and discharges—of the various production runs using output obtained from approximately the thirtieth iteration of each run?

Some Seminal Results of Running the Model

The combinations of decision variables used in selected runs of the model are indicated in table 36, along with the number of iterations for each run and the resulting objective function values. The objective function values represent the additional costs ($ million per day) over base costs, associated with improving the ambient environmental quality of the region —atmosphere, estuary, and landfills. The base case is the situation that existed in the Lower Delaware Valley during the period of roughly 1968 to 1970. The base cost was determined by using the Lower Delaware Valley REQM model with landfills of low quality throughout the region, and with only the physical restrictions on liquid and gaseous residuals discharges which existed in the 1968–1970 period. Thus the objective function value for the base case is zero.

The results of analyses illustrate: (1) total regional costs to meet various combinations of ambient air, ambient water, and landfill quality standards; (2) the effects of including regional facilities as components of regional REQM; (3) linkages among forms of residuals and among the three environmental media; and (4) distributions of changes in ambient environmental quality and of REQM costs. The first three are illustrated in figures 25, 26, 27, 28, and table 37, and associated text.

Costs and Environmental Quality Standards

Figure 25 shows the total increased costs to the region of meeting: alternative air quality standards; alternative water quality standards; and alternative air and water quality standards simultaneously.

As expected, total costs to the region increase as the air and water quality standards become more stringent. Note that total costs to the

---

[13] The one exception to the total discharges not changing much beyond the thirtieth iteration is the tradeoff between instream aerators and BOD_5 discharges (and perhaps other discharges).

TABLE 36. Characterization of Selected Production Runs of the Lower Delaware Valley Model and Resulting Values of the Objective Function

Run number	Standards[a]			Regional options[b]	Iterations of regional model	Objective function value[c] 10^6 \$/day
	Air	Water	Landfill			
1	T	T	H	+	34	1.20
2	T	E	H	+	35	1.18
3	E	T	H	+	29	0.42
4	E	E	H	+	30	0.36
5	T	T	M	+	35	1.16
6	T	E	M	+	34	1.14
7	E	T	M	+	30	0.37
8	E	E	M	+	30	0.31
9	T	O	H	+	35	1.12
10	O	E	H	+	30	0.11
11	O	T	H	+	16	0.14
12	E	O	H	+	30	0.26
13	E	T	H	O	30	0.52
14	E	E	H	O	30	0.35
15	E	T	H	O[d]	29	0.46
16	E	T	H	+[d]	30	0.39
17[e]	∞	O	H	O	1	1.19
18[e]	O	∞	H	O	1	0.34
19[e]	∞	∞	H	O	1	1.41
23	O	∞	H	+	1	0.36
25	O	O	H	O	1	0.034
26	O	O	M	O	1	0.016
Base Case	O	O	L	O	1	0

Source: Spofford, Russell, and Kelly, *Environmental Quality*, p. 81.

[a] Standards are *Air and water quality:* T = tight, E = easy, O = no standards, ∞ = extremely high marginal penalties on residuals discharges; *Landfill quality:* H = high, M = medium, L = low. (See table 35.)

[b] The regional options include: two regional paper recycling plants (for newsprint and paperboard), two regional sewage treatment plants, and instream aeration. The options of using used newspapers as input to the newsprint plant and used corrugated containers as input into the linerboard plant as alternatives for reducing solid residuals management costs are available in all runs. The regional sewage treatment plant option is available only in runs 15 and 16. + means instream aeration is employed in the run; O means instream aeration is not employed in the run.

[c] Increased costs are net of base costs.

[d] Regional sewage treatment plant option is also employed.

[e] Runs 20, 21, 22 are similar to runs 17, 18, and 19, respectively, except that medium quality landfills rather than high quality landfills are specified.

region of meeting T-level water quality standards in the absence of any constraints on the discharge of gaseous residuals is substantially less than the costs of meeting T-level air quality standards in the absence of any constraints on the discharge of liquid residuals. This is due in part to the fact that most of the gaseous residuals dischargers in the region have been taken into account in the model, but only those liquid residuals dischargers which discharge their waterborne residuals to the Delaware

<table>
<tr><td colspan="5" align="center">Air quality standards
($ million per day)</td></tr>
</table>

		O	E	T	∞
Water quality standards ($ million per day)	O	0.034[a] (#25)	0.26 (#12)	1.12 (#9)	1.19[a] (#17)
	E	0.11 (#10)	0.36 (#4)	1.18 (#2)	
	T	0.14 (#11)	0.42 (#3)	1.20 (#1)	
	∞	0.34[a] (#18)			1.41[a] (#19)

Figure 25. Total increased costs to the Lower Delaware Valley region of meeting different sets of air and water quality standards, with high quality landfills. Note: numbers in parentheses are production run numbers. For meaning of O, E, T, and ∞, see table 35 and footnotes to table 36. From: Spofford, Russell, and Kelly, *Environmental Quality,* p. 60.

[a] Single iteration runs, the outputs of which are not dependent on the non-linear optimization algorithm. Aerators are not employed in these runs.

Estuary have been considered. Nevertheless, it appears that "cleaning up" the air will be a more pervasive and costly problem for the region than "cleaning up" the water.[14]

Figure 26 illustrates the effects of different landfill qualities in conjunction with different sets of ambient air and water quality standards. The total incremental costs to the region of going from medium quality landfills to high quality landfills amount to between $40 thousand and $50 thousand per day, depending on the levels of air and water quality standards. However, as indicated by the module 5 costs, only $25 thousand to $30 thousand per day are attributable to the paper plant-solid residuals management module. Thus, landfill quality requirements impose substantial incremental costs on other, apparently unrelated, activities in the region. A second point to note from figure 26 is that the water quality standards seem to have very little impact on the costs of paper production from paper residuals and on the costs of solid residuals management

[14] This is particularly true, because it was assumed, at the time the model inputs were developed, that natural gas was available to all users at a relatively low price.

Total regional costs:
($ million per day)

		Air quality standards				Air quality standards				Air quality standards	
		E	T			E	T			E	T
Water quality standards	E	0.36 (#4)	1.18 (#2)		E	0.31 (#8)	1.14 (#6)		E	0.05	0.04
	T	0.42 (#3)	1.20 (#1)		T	0.37 (#7)	1.16 (#5)		T	0.05	0.04
		High quality landfills				Medium quality landfills				Difference in costs	

Paper plant–solid residuals management (module 5) costs:[a]
($ million per day)

		Air quality standards				Air quality standards				Air quality standards	
		E	T			E	T			E	T
Water quality standards	E	0.060 (#4)	0.079 #(2)		E	0.038 (#8)	0.054 #(6)		E	0.022	0.025
	T	0.071 (#3)	0.079 (#1)		T	0.040 (#7)	0.055 (#5)		T	0.031	0.024
		High quality landfills				Medium quality landfills				Difference in costs	

Figure 26. Total increased costs to the Lower Delaware Valley region, with high and medium quality landfills. Note: For meaning of O, E, T, and ∞, see table 35 and footnotes to table 36. Numbers in parentheses are production run numbers. From: Spofford, Russell, and Kelly, *Environmental Quality,* p. 66.

[a] Costs of paper production from paper residuals plus costs of other solid residuals handling and disposal.

(module 5). However, the air quality standards appear to have a substantial impact on these costs.

Costs and Regional Facilities

The effects on total REQM costs of including regional options of instream aeration and regional sewage treatment plants are illustrated by the results

Air quality standards
(tons per day)

		O	E	T	∞
Water quality standards (tons per day)	O	550 (#25)	500 (#12)	0.0 (#9)	0 (#17)
	E	160 (#10)	155 (#4)	165 (#2)	
	T	0.0 (#11)	0.0 (#13)	0.0 (#1)	
	∞	0 (#18)			0 (#19)

Figure 27. Quantities of used newspaper input into new newsprint plant under varying environmental quality standards, with high quality landfills. Note: Numbers in parentheses are production run numbers. For meaning of O, E, T, and ∞, see table 35 and footnotes of table 36. From: Spofford, Russell, and Kelly, *Environmental Quality,* p. 113.

of four runs—3, 13, 15, 16. These runs all have T-level water quality standards, E-level air quality standards, and high quality landfills. When neither the instream aeration nor regional sewage treatment plant options is included, total regional costs are $520 thousand per day (run 13); with regional sewage treatment but no instream aeration, total regional costs amount to $460 thousand per day (run 15); with instream aeration but no regional treatment, costs are $420 thousand per day (run 3); and with both regional treatment and instream aeration, total regional costs are $390 thousand per day (run 16). Thus, the inclusion of regional facilities for residuals management can significantly decrease total regional costs to achieve specified levels of ambient environmental quality.

The input activity level of a regional newspaper plant illustrates how REQM conditions affect the economic feasibility of recycling used paper products in the region. Figure 27 shows the inputs to a "new" newsprint plant under varying AEQ standards. From this, two conclusions can be drawn. First, operation of the new newsprint plant is very sensitive to landfill costs, as would be expected. For low landfill quality coupled with no AEQ standards (base case), the input to the plant is zero. However, for high landfill costs but still no AEQ standards, the input is 550 tons per day (run 25). Under these conditions, substantial amounts of used newspapers are collected from single-family residences throughout the region, mostly in Philadelphia, in addition to the amounts currently being collected for recycling. Second, the new newsprint plant is sensitive to higher levels of both ambient water and air quality standards, that is, activity level reduces to zero for all combinations above E-E, except for E-water combined with T-air, because the costs of residuals management at the newsprint plant using used newspapers increase rapidly at higher AEQ standards.

Linkages

One of the original objectives of applying the REQM modeling framework to the Lower Delaware Valley region was to investigate the linkages among gaseous, liquid, solid, and energy residuals, and among the various environmental media, in an existing situation. The evidence of linkages among the forms of residuals and among the three environmental media may be obtained from a comparison of: objective function values; magnitudes of materials *flows* in the region, such as fuels, residuals discharges, residuals transport (mixed solid residuals and sludge barging), and used paper products; capacities of *structures* such as municipal incinerators, regional sewage treatment plants, and add-on residuals modification devices (for example, electrostatic precipitators); or both.

Suggested evidence of linkages is presented in figure 28. In part A of that figure, objective function values for the region are shown for varying combinations of ambient air and water quality standards, with high quality landfills. With existing ambient air quality standards (that is, the O level), it costs the region about $76 thousand per day to go from existing conditions to E-level water quality standards. When E-level air quality standards are imposed, the costs of improving water quality to the E-level are about $100 thousand per day, a difference of about $24 thousand. Had there been no linkages at all between air and water, the increased costs would have been the same for both cases. In going from E-level water quality to T-level, the difference is larger: about $30 thousand per day when air quality is not considered, compared with about $60 thousand per day when the E-level air quality standard is imposed, a difference of about $30 thousand.

Part A

Total regional costs | *Incremental regional costs*
(\$ million per day) | (\$ millon per day)

Air quality standards

		Total regional costs		Incremental regional costs	
		O	E	O	E
Water quality standards	O	0.034 (#25)	0.26 (#12)		
				0.076	0.10
	E	0.11 (#10)	0.36 (#4)		
				0.03	0.06
	T	0.14 (#11)	0.42 (#3)		

Part B

Total MPSX 1 costs | *Incremental MPSX 1 costs*
(\$ million per day) | (\$ million per day)

Air quality standards

		Total MPSX 1 costs		Incremental MPSX 1 costs	
		O	E	O	E
Water quality standards	O	0.0 (#25)	0.11 (#12)		
				0.020	0.04
	E	0.020 (#10)	0.15 (#4)		
				0.004	0.02
	T	0.024 (#11)	0.17 (#3)		

Figure 28. Evidence of linkages among forms of residuals: Lower Delaware Valley model, with high quality landfills. Note: Numbers in parentheses are production run numbers. For meaning of O, E, T, and ∞, see table 35 and footnotes to table 36. From: Spofford, Russell, and Kelly, *Environmental Quality,* pp. 110–111.

In part B, evidence is based on the costs from the first module (MPSX 1) containing the petroleum refineries, the steel mills, and the electric power plants. When no air quality standards are imposed, the cost to move from O-level to E-level water quality standards is about $20 thousand per day. When E-level air quality standards are imposed simultaneously, the incremental costs to achieve E-level water quality standards are about $40 thousand per day. In moving from E-level to T-level water quality standards when air quality standards are not imposed, the incremental cost is about $4 thousand per day. When E-level air quality standards are imposed simultaneously with T-level water quality standards, the incremental cost is about $20 thousand per day. Again, the linkages between air quality and water quality in the region are clearly indicated.

The final evidence of linkages presented here is the interrelationship between different levels of landfill quality and ambient air and water quality standards. Data are presented for two cases in table 37, one for E-level air and water quality standards, and one for T-level air and water quality standards. As both panels of the table show, if there were no linkages between landfill quality and air and water quality, the difference between total regional costs (column 1) and the difference between solid residuals management costs (column 2) would have been the same. That they are substantially different indicates that linkages exist. Thus, substantial costs are imposed on other sectors when higher quality landfills are required, the linkages to these additional costs being through the air and water quality of the region.

Distributional Effects

With respect to the distribution of ambient environmental quality, table 38 shows the distribution of SO_2 concentrations among the fifty-seven political jurisdictions for runs 1 and 3, the T-level air quality standards and the E-level air quality standards, respectively. The maximum concentration for both runs is found in political jurisdiction 16 (South Philadelphia). Of the ten highest concentrations for run 1, eight are located within Philadelphia (jurisdictions 14, 15, 16, 17, 18, 19, 23, and 27), one in Delaware County (jurisdiction 9), and one in Camden, New Jersey (jurisdiction 50). Although not shown in this table, the TSP distribution for run 1 is similar to that for SO_2 in that of the ten highest concentrations, nine are located within Philadelphia (jurisdictions 14,

TABLE 37. Relationship Between Solid Residuals Management and Air and Water Quality Management: Lower Delaware Valley Model ($ million per day)

Panel A

With easy (E) air and water quality standards

Quality of landfill	Total costs to region	Solid residuals management costs (MPSX 5)
High quality landfill (#4)	0.36	0.060
Medium quality landfill (#8)	0.31	0.038
Difference	0.05	0.022

Panel B

With tight (T) air and water quality standards

Quality of landfill	Total costs to region	Solid residuals management costs (MPSX 5)
High quality landfill (#5)	1.20	0.079
Medium quality landfill (#5)	1.16	0.055
Difference	0.04	0.024

Note: Numbers in parentheses are production run numbers.
Source: Spofford, Russell, and Kelly, *Environmental Quality,* p. 57.

15, 16, 19, 22, 23, 24, 26, 27) and one in Camden (jurisdiction 50). Unlike SO_2, however, the maximum concentration of TSP for runs 1 and 3 is found in jurisdiction 15 (in Center City Philadelphia). These results demonstrate that, for virtually all reasonable REQM strategies, the poorest air quality throughout the region, with respect to SO_2 and TSP concentrations, will be found in the Philadelphia-Camden area.

The distributions of dissolved oxygen (DO) concentrations throughout the twenty-two-reach estuary for runs 1, 2, 3, 13, 15, and 16 are shown in table 39. When E-level water quality standards are imposed (run 2), the eight reaches with the poorest water quality (which actually is fairly high quality), are all located between reaches eleven and nineteen, from the confluence with the Schuylkill River to below Wilmington. Using DO as an indicator of water quality, estuary water quality upstream from the confluence with the Schuylkill River is better than downstream. The reach with the poorest water quality is 17, near Wilmington, Delaware.

For the T-level water quality standard (DO $\geq$ 5.9 mg/l), the distributions of the reaches with poorer quality change (runs 1, 3, 13, 15, and

TABLE 38. Distribution of Sulfur Dioxide Concentration by Political Jurisdiction for Selected Production Runs, with High Quality Landfills (micrograms per cubic meter)

Political jurisdiction	Sulfur dioxide		Political jurisdiction	Sulfur dioxide	
	Run 1 T-level ambient air & water qual. stds.	Run 3 E-level ambient air & T-level ambient water qual. stds.		Run 1 T-level ambient air & water qual. stds.	Run 3 E-level ambient air & T-level ambient water qual. stds.
1	17	54	30	23	66
2	15	49	31	16	46
3	6	14	32	13	44
4	12	39	33	12	40
5	8	22	34	11	35
6	5	19	35	11	32
7	5	12	36	6	20
8	22	53	37	12	37
9	32[a]	68	38	9	27
10	25	58	39	4	12
11	28	64	40	11	46
12	15	42	41	9	34
13	11	32	42	7	24
14	56[a]	96[a]	43	4	11
15	42[a]	98[a]	44	13	40
16	83[a,b]	120[a,b]	45	10	34
17	64[a]	100[a]	46	8	20
18	35[a]	70	47	12	49
19	50[a]	91[a]	48	11	40
20	24	57	49	8	25
21	18	48	50	40[a]	87[a]
22	29	74[a]	51	20	64
23	38[a]	89[a]	52	27	71
24	29	103[a]	53	16	43
25	19	66	54	6	18
26	25	70	55	22	51
27	34[a]	94[a]	56	9	26
28	25	65	57	8	26
29	17	49			

Note: For locations of political jurisdictions, see figures 23 and 24.

Source: Bower, *Regional Residuals*, p. 94.

[a] Indicates the ten highest concentrations of SO_2.

[b] Indicates the minimum concentration of SO_2.

TABLE 39. Distribution of Dissolved Oxygen Concentration by Estuary Reach for Selected Production Runs, with High Quality Landfills (milligrams per liter)

Delaware Estuary reach	Dissolved oxygen					
	Run 1[a]	Run 2[b]	Run 3[c]	Run 13[c]	Run 15[c]	Run 16[c]
1	8.6	8.6	8.6	8.6	8.6	8.6
2	7.7	7.7	8.0	7.9	7.9	7.9
3	6.9	6.6	7.2	7.0	6.9	7.2
4	6.3	6.0	6.8	6.2	6.1	6.5
5	5.9[d]	5.7	6.4	5.8[d]	5.4[d]	6.0
6	6.0	5.9	6.5	6.0	5.4[d]	6.0
7	5.9[d]	5.9	6.3	5.8[d]	5.3[d]	6.0
8	5.9	5.8	6.3	6.2	5.7	5.8[d]
9	6.4	6.1	6.7	6.8	6.3	6.2
10	6.8	5.3	7.0	7.8	7.5	7.1
11	5.3[d]	3.6[d]	5.1[d]	5.7[d]	5.2[d]	5.1[d]
12	6.1	3.7[d]	5.7[d]	6.3	6.0	6.1
13	5.7[d]	3.6[d]	5.4[d]	5.7[d]	5.3[d]	5.0[d]
14	5.7[d]	4.0[d]	5.5[d]	5.7[d]	5.3[d]	5.1[d]
15	6.1	4.5[d]	6.0[d]	6.0	5.5[d]	5.8[d]
16	6.4	5.2	6.3	6.0[d]	5.6	6.4
17	5.0[d,e]	3.0[d,e]	5.0[d,e]	4.9[d,e]	5.0[d,e]	5.0[d,e]
18	5.1[d]	3.4[d]	5.2[d]	5.4[d]	5.5	5.1[d]
19	5.7[d]	4.8[d]	5.8[d]	6.0	6.1	5.8[d]
20	6.1	5.8	6.1	6.1	6.2	6.1
21	6.2	6.2	6.3	6.3	6.3	6.3
22	6.6	6.6	6.6	6.7	6.7	6.6

Note: For location of reaches, see figure 21.

Source: Bower, *Regional Residuals*, p. 95.

[a] T-level ambient water and air quality standards.

[b] E-level ambient water and T-level ambient air quality standards.

[c] T-level ambient water quality and E-level ambient air quality standards.

[d] Indicates the eight lowest concentrations of dissolved oxygen.

[e] Indicates the minimum concentration of dissolved oxygen.

16). Now the eight reaches with poorer water quality are distributed between reaches 5 and 19, from above Philadelphia to below Wilmington, but the distributions differ depending on whether or not regional sewage treatment plants or instream aeration or both are employed. As before, the reach with the poorest water quality is 17, near Wilmington. As with the E-level standards, the poorest quality water is not found in the vicinity of Philadelphia.

With respect to the distribution of costs, table 40 shows for each of the fifty-seven political jurisdictions:

1. the percentage increase in household heating costs with
   —T-level air and T-level water quality standards and high quality landfills (run 1), and
   —E-level air and T-level water quality standards and high quality landfills (run 3)
2. the percentage increase in household electricity costs under the same two conditions as above
3. the percentage increase in municipal solid residuals management costs for
   —high quality landfills with E-level air and T-level water quality standards (run 3), and
   —medium quality landfills with E-level air and T-level water quality standards (run 7)
4. the increase in costs of sewage disposal per sewered household under
   —T-level water and T-level air quality standards with high quality landfills (run 1), and
   —E-level water and T-level air quality standards (run 2).

With respect to household heating costs, for the T-level air quality standards (run 1), the largest increase in costs, about 55 percent, occurs in jurisdiction 23, in central Philadelphia. The next largest increase, about 50 percent, occurs in an adjacent jurisdiction, number 19, also in Philadelphia. For the E-level air quality standards (run 3), the largest and next largest increases, about 30 percent and 28 percent, respectively, also occur in jurisdictions 23 and 19, respectively.

As for increases in household electricity costs, a different distribution exists. For the T-level air quality standards (run 1), the largest increase in costs, about 16 percent, occurs in New Castle, Delaware (jurisdictions 1–4). The next largest increase, about 14 percent, occurs in Chester, Delaware, Philadelphia, Montgomery, and Bucks counties in Pennsylvania (jurisdictions 5–42). In contrast, in Camden, New Jersey (jurisdictions 50–54), where the air quality is almost as poor as in Philadelphia, the increase in household electricity costs amounts to only about 3 percent. For the E-level air quality standards (run 3), the situation reverses, and the Pennsylvania counties, including Philadelphia (juris-

TABLE 40. Distribution of Increased Consumer Costs by Political Jurisdiction for Selected Production Runs

Political juris-diction	Household heating		Household electricity[b]		Municipal solid residuals management[b]		Municipal sewage	
	Run 1[a]	Run 3	Run 1	Run 3	Run 3	Run 7	Run 1	Run 2
	(percentage)						($/household/ year)[e]	
1	25.1	1.8	15.8[d]	2.4	19.2	7.7	0.5[e]	0.1
2	30.5	0.7	15.8[d]	2.4	19.4	7.8	43.5	8.0
3	29.0	0[e]	15.8[d]	2.4	19.3	7.7	5.9	1.1
4	23.0	0[e]	15.8[d]	2.4	19.4	7.8	1.0	0.2
5	33.9	0.6	13.6	4.5[d]	19.1	7.6	—	—
6	35.5	0[e]	13.6	4.5	19.1	7.7	—	—
7	16.8	0[e]	13.6	4.5	19.4	7.8	—	—
8	26.1	9.6	13.6	4.5	10.8	7.7	3.3	2.5
9	23.3	7.2	13.6	4.5	17.9	7.7	2.1	1.0
10	24.5	3.2	13.6	4.5	19.3	7.7	3.5	1.9
11	28.3	0.9	13.6	4.5	10.9	7.7	16.2	8.0
12	27.8	3.4	13.6	4.5	18.8	7.7	3.9	2.1
13	26.2	1.9	13.6	4.5	21.0	7.7	2.5	1.7
14	19.7	6.1	13.6	4.5	8.7	7.7	21.4	16.1
15	41.0	25.8	13.6	4.5	7.6[e]	7.6	16.2	12.2
16	31.4	18.5	13.6	4.5	17.3	7.7	15.4	10.6
17	16.4	7.4	13.6	4.5	8.1	7.7	19.3	13.1
18	21.7	8.2	13.6	4.5	17.5	7.7	18.8	12.8
19	50.4	28.5	13.6	4.5	20.1	7.6	14.4	9.8
20	22.2	6.3	13.6	4.5	20.4	7.7	16.8	11.4
21	25.6	6.3	13.6	4.5	8.6	7.6	16.6	11.6
22	31.7	15.2	13.6	4.5	20.4[d]	7.7	19.3	14.0
23	55.4[d]	30.0[d]	13.6	4.5	11.2	7.5	14.8	10.6
24	14.1	0.2	13.6	4.5	17.6	6.6[e]	8.0	7.1
25	10.4	0.5	13.6	4.5	17.8	6.6[e]	16.0	14.2
26	12.6	0.9	13.6	4.5	23.1	8.2[d]	15.0	13.4
27	25.5	4.8	13.6	4.5	20.5	7.7	22.1	16.7
28	29.8	5.1	13.6	4.5	20.4	7.7	15.6	13.3
29	33.9	1.6	13.6	4.5	10.7	7.6	17.9	14.1
30	16.5	2.7	13.6	4.5	20.6	7.7	15.6	13.9
31	17.2	2.7	13.6	4.5	20.3	7.7	15.7	13.6
32	9.9[e]	0.6	13.6	4.5	20.2	7.6	15.8	14.0
33	31.4	1.1	13.6	4.5	20.4	7.7	19.7	17.6
34	35.1	1.9	13.6	4.5	18.7	7.5	7.3	4.9
35	26.9	2.8	13.6	4.5	19.1	7.7	2.3	1.5
36	31.2	0.0	13.6	4.5	19.2	7.7	—	—
37	26.7	0.7	13.6	4.5	19.0	7.6	6.1	5.4
38	26.9	0.6	13.6	4.5	19.0	7.6	1.4	1.2
39	36.7	0.0[e]	n.a.	n.a.	19.2	7.7	—	—

(continued)

235

TABLE 40. (continued)

Political juris-diction	Household heating Run 1[a]	Household heating Run 3	Household electricity[b] Run 1	Household electricity[b] Run 3	Municipal solid residuals management[b] Run 3	Municipal solid residuals management[b] Run 7	Municipal sewage Run 1	Municipal sewage Run 2
			(percentage)				($/household/year)[c]	
40	36.7	0.2	13.6	4.5	19.1	7.6	8.3	1.3
41	31.3	0.0[e]	13.6	4.5	19.2	7.7	15.9	0.0[e]
42	29.6	0.3	13.6	4.5	19.3	7.7	1.8	1.6
43	33.1	0.0[e]	6.8	2.2	19.3	7.7	—	—
44	34.9	0.0[e]	2.9	0.4	19.4	7.7	48.2[d]	1.2
45	24.2	0.0[e]	2.9	0.4	19.2	7.7	—	—
46	25.0	0.0[e]	2.9	0.4	19.2	7.7	—	—
47	25.1	0.8	2.9	0.4	19.3	7.7	12.7	3.0
48	17.4	0.0[e]	2.9	0.4	19.5	7.8	11.5	0.0[e]
49	19.7	0.0[e]	1.4[e]	0.2[e]	19.4	7.8	2.9	0.0[e]
50	31.7	9.2	2.9	0.4	19.5	7.8	47.8	36.9[d]
51	20.6	4.1	2.9	0.4	19.3	7.7	20.7	13.6
52	29.5	6.0	2.9	0.4	19.4	7.8	4.4	3.2
53	22.3	2.8	2.9	0.4	19.4	7.8	—	—
54	25.6	0.7	10.6	1.3	19.3	7.7	—	—
55	25.9	5.0	2.9	0.4	19.5	7.8	6.6	4.5
56	19.2	0.2	10.6	1.3	19.5	7.8	—	—
57	28.2	0.0	10.6	1.3	19.6	7.9	0.5	0.1

Notes: For locations of political jurisdictions, see figures 23 and 24. Dashes = not applicable. Standards for the runs are:

Run	Air	Water	Landfills
1	T	T	H
3	E	T	H
7	E	T	M
2	T	E	H

Source: Bower, *Regional Residuals*, p. 97.

[a] Cost increases for household heating reflect the technological upper limits except in jurisdictions 1, 2, 7, 12, 35, 49, and 56. There is empirical evidence to suggest that at the "optimum" for this run, all cost increases would be at their upper limits. Cost increases at the upper limits represent a situation where there is total conversion within the region to natural gas.

[b] The situation in which many jurisdictions have the same percentage cost increase reflects the fact that multiple jurisdictions are served by the same utility or solid residuals management agency.

[c] The costs of sewage disposal have been allocated in the model on the basis of both flow and BOD_5 load. BOD_5 concentrations in sewage have been assumed using different loading factors for urban, suburban, and rural areas. The average increased cost to jurisdictions for sewage disposal is computed on the basis of total sewage disposal costs divided by total housing units for each jurisdiction. Thus, all other factors the same, jurisdictions that are less than 100 percent sewered will register smaller average increases than those that are 100 percent sewered. This accounts for the apparent discrepancy in increased costs for jurisdictions served by the same sewage treatment plant.

[d] Maximum increases.

[e] Minimum increases.

dictions 5–42), are associated with the largest increase in costs, about 4.5 percent.

Different distributions among political jurisdictions of increases in household heating and household electricity costs reflect the fact that household heating is achieved largely by energy conversion on the site, whereas household electricity is produced in a few large plants in the region. The distributions are also affected by the location of each generator in relation to the receptors in the region, for example, the number of receptors that are downwind of the discharge. The electrical generating stations in New Castle County, Delaware, are upwind of much of the rest of the region; hence, their discharges have widespread effects.

There is relatively little variation among the jurisdictions in the increased costs of municipal solid residuals management. This is because, for all runs, all jurisdictions were required to maintain the same quality landfills, that is, high, medium, or low quality. Where a significant variation does occur, it is associated with the high landfill quality runs, such as run 3. Within this group of runs, the differences in increased costs among jurisdictions for the E-level air quality standards are substantially greater than those associated with the T-level standards. This is because, for the E-level air quality standards, seven municipal incinerators (out of a total of twenty-three), all available in Philadelphia, are used for solid residuals management. The result of this is that the solid residuals management costs of the six jurisdictions within Philadelphia that use these incinerators are approximately 50 percent of those jurisdictions that do not have this option. However, when T-level air quality standards are imposed, some of the incinerators are not used, resulting in more uniform cost increases throughout the region.

With respect to municipal sewage disposal costs, among the ten jurisdictions exhibiting the largest average increased costs in terms of dollars per household per year, are Wilmington (jurisdiction 2), Trenton (jurisdiction 44), and Camden (jurisdiction 50). In fact, one of these three jurisdictions is always in the number one position—with the largest increase in costs of municipal sewage disposal. The remaining seven of the ten highest cost jurisdictions are located within the city of Philadelphia.

Caveats and Conclusions

Inevitably, in a research undertaking as large and complicated as that of developing and applying a regional REQM model to an existing complex region, limitations will be imposed on the analysis and difficulties will be encountered. This final section of the chapter consists of four parts: (1) limitations presented as "some caveats," (2) conclusions relating specifically to REQM for the Lower Delaware Valley region; (3) some difficulties encountered, and (4) conclusions relating to methodological issues in analysis for REQM. But before proceeding it should be noted that from the beginning the application was intended to be illustrative and not operational, although the model is based on actual data from the region and on the best data available at the time the model was developed.

Some Caveats

1. The model is largely based on publicly available data of unknown quality and has only been partially calibrated. No field measurements of any type were made by the modeling team, and the implications of the results presented here have not been assessed in the field.

2. The model reflects conditions that existed in the region about 1970. For example, fuel availability and costs in the model do not reflect recent increases resulting from worldwide changes in petroleum prices.

3. To keep the model as small as possible, all stacks at the same x-y location in EPA's inventory of discharges of gaseous residuals were aggregated. This aggregation procedure could have a substantial effect on the accuracy of predicted ground level concentrations, especially near large, multistack sources.

4. Some of the important industries in the region—particularly the petroleum refineries and the steel mills—are represented in the model by "collapsed" versions of larger, generalized models of industrial operations. These collapsed versions were calibrated to the production levels of specific plants, but other, perhaps more important details peculiar to specific plants have not been included.[15]

5. The receptor locations chosen to represent ambient air quality over the fifty-seven political jurisdictions were not selected with respect to existing locations of residential areas or with respect to what might be critical locations within each of the jurisdictions. They were simply chosen to be roughly in the central portion of each jurisdiction.

6. Certain categories of residuals generators and dischargers, such as transportation, have not been incorporated in the model. Nonpoint

[15] See the discussion of this point in footnotes 46 and 47 on pages 99 and 100 of Spofford, Russell, and Kelly, *Environmental Quality*.

sources of residuals discharges, such as agricultural and construction activities, have been included only as they affect background concentrations.

7. The costs reflected in the model are based on meeting standards expressed in terms of long-term average ambient concentrations (annual averages for air quality and low flow summer averages for estuary water quality), rather than on meeting shorter term peak concentration constraints. The impact on REQM strategies, REQM costs, and distribution of REQM costs of the latter consideration have not been assessed.

8. The predictions of ambient air quality for the region are based on the Implementation Planning Program's (IPP) dispersion model, on 1970 meteorological data for Philadelphia, and on the associated calibration relationships, all supplied by the U.S. Environmental Protection Agency. The results clearly depend heavily on these data and estimated relationships. However, time and the resources available precluded the possibility of investigating the sensitivity of the results either to alternative dispersion models or to alternative meteorological data.

9. The results reflect two optional physical measures for improving ambient air quality in the region that may not be realistic. First, complete conversion to natural gas in household and commercial heating activities is allowed in the regional model; it is also allowed in the electric power plants and some of the industrial plants. Second, barging of municipal sewage sludge to sea is allowed for the larger cities along the estuary.

10. Imports of gaseous residuals into the Lower Delaware Valley region are included only in terms of background concentrations. No limitations were imposed on the export of such residuals from the region. Given the increasing knowledge about the long-range transport of gaseous residuals, establishing constraints on the flows of gaseous residuals across the boundary of the region in the direction of the prevailing winds might have resulted in additional REQM costs to the region.

11. The REQM costs estimated in the model reflect actual resource costs to society for achieving the associated AEQ standards; they do not reflect the actual cash flows that would be incurred by the individual dischargers. The latter are a function of the multiple and complicated subsidies and tax structures at federal, state, and local levels (such as federal grants to municipalities for sewage treatment plant construction and federal tax provisions allowing individual, industrial, and other operations to take rapid depreciation for investments in approved pollution control equipment). Thus, developing information on distribution

of actual out-of-pocket costs to different dischargers would be even more difficult than the cost distribution information generated in this study.

Some Conclusions: REQM for the Lower Delaware Valley Region

1. All cases explored show: (A) substantial variation in air quality, in terms of SO_2 and TSP concentrations, over the Lower Delaware Valley region; and (B) substantial variation in water quality, in terms of dissolved oxygen, algae, and fish biomass concentrations, throughout the Delaware Estuary. In general, the lowest concentrations of SO_2 and TSP are in Bucks, Chester, and Montgomery counties in Pennsylvania; the highest in Philadelphia County in Pennsylvania and Camden County in New Jersey. With respect to dissolved oxygen, the highest dissolved oxygen levels are always found in reach 1 near Trenton, New Jersey; the lowest in reach 17, near Wilmington, Delaware.

2. Air quality over the Lower Delaware Valley region appears to be more of a problem than Delaware Estuary water quality for at least two reasons. First, in some areas of Philadelphia, it appears that it may be technologically difficult to achieve, and then to maintain, the federal primary annual average standards for SO_2 and TSP. Second, regional costs to achieve the federal primary ambient air quality standards would be six to eight times the regional costs of achieving ambient water quality standards in the estuary, that is, between $300 and $400 million per year versus about $50 million per year, respectively.

3. Ambient water quality standards in the estuary can be met in all cases examined by using various combinations of: on-site residuals discharge reduction, regional sewage treatment, and instream aeration. The regional alternatives appear to allow the region as a whole to meet the same ambient water quality standards at less cost than by relying on residuals discharge reduction measures at individual sites. However, all dischargers do not benefit to the same degree from the decreased cost. Indeed, some dischargers who are not serviced by the regional facilities actually pay more under management alternatives with regional facilities than when management alternatives are restricted to measures at individual sites and local municipal sewage treatment.

4. Under the conditions reflected in the model, roughly those of 1970, the rail haul alternative for disposing of solid residuals appears to be too expensive. However, as landfill sites become scarce (and hence more

expensive), this alternative will certainly become more attractive, as will increased recycling of used paper products.

5. Again, under 1970 conditions, it appears that new municipal incinerators are not economically feasible alternatives for solid residuals management, especially with the stringent standards that would be imposed on discharges of liquid and gaseous residuals from them.

6. Based on the analyses presented, there exists a variety of combinations of physical measures for meeting the regional AEQ standards, and these combinations do not differ very much in total costs to the region. However, the distributions of levels of residuals discharge reduction, of residuals discharges, and of costs among dischargers, all vary substantially from combination to combination.

Some Difficulties

Space limitations do not allow a detailed discussion of all the difficulties encountered in the development of a model for regional REQM analysis and its application to the Lower Delaware Valley region.[16] However, listing some difficulties along with some conclusions about methodology should provide useful insights.

1. Clearly, an analysis of a region as complicated as the Lower Delaware Valley requires large amounts of specific data, such as data on types and quantities of residuals discharges from individual point and nonpoint sources and costs of reducing such discharges to various degrees. Only a limited amount of such data was readily available, particularly with respect to costs for *individual* dischargers (in contrast to generalized cost functions for some typical processes used to modify liquid residuals, such as primary sedimentation and trickling filters). Similarly, relatively few actual observations of the biological aspects of ambient water quality associated with observed sets of residuals discharges were available. This meant that, for example, even if a model of the aquatic ecosystem were constructed from first principles and published laboratory relationships, its calibration and verification would be difficult, and perhaps even impossible, given the lack of resources for primary data collection in the field.

---

[16] Various issues with respect to analysis for regional REQM are explored in Bower, *Regional Residuals.*

2. Even if all the necessary data on both residuals discharges and ambient environmental quality had been available, the size of even the simplest model of the complex number, type, and distribution of activities in the region would be such as to involve significant computational costs. Computer costs are directly related to model size, which is in turn a function of: (A) the number of activities in the region explicitly modeled; (B) the incorporation of nontreatment residuals discharge reduction alternatives; (C) the explicit consideration of the linkages both among liquid, gaseous, and solid and energy residuals, and among the three environmental media (atmosphere, watercourses, and land); and (D) the provision of information on the distributions of ambient environmental quality and of REQM costs. Thus, in order to do analysis at all, ways had to be developed of aggregating data while minimizing the loss of important information.

3. The decision to incorporate a nonlinear aquatic ecosystem model of the estuary required the use of an heuristic, iterative nonlinear programming algorithm. This in turn resulted in potential problems of multiple optima and problems of determining the steady state values of the biological state variables in the aquatic model. To use an heuristic, iterative nonlinear programming algorithm efficiently requires the development of an efficient step size selector.

4. The size of the model and the use of an iterative nonlinear programming algorithm raised the problem of providing comparable outputs from one run to another. As noted previously, in order to save on computer expenditures, all runs in which the nonlinear algorithm was used were arbitrarily stopped near the thirtieth iteration, short of even a local optimum. Unfortunately, it appears that the rate at which the algorithm converges to an optimum is dependent on the levels of AEQ standards specified, among other factors. For example, the algorithm appears to converge more slowly for the runs with stringent ambient air quality standards than for those with the less restrictive air quality standards. Thus, what is not known is exactly which iteration of one computer run would be "comparable" to which iteration of another computer run, and under what conditions. However, large differences between costs and discharges near the thirtieth iteration of the different production runs *can* be compared and used for evaluation of alternatives; it is the small differences about which strong statements cannot be made.

5. Generating information on the distribution of costs involved in achieving specified levels of ambient environmental quality was straight-

forward. However, real difficulty arose when an attempt was made to constrain simultaneously the levels of ambient environmental quality to be achieved *and* the levels of consumer cost increases.

6. The substantial volume of output information generated by such a model for such a complicated region—particularly when the distributional aspects are included—represents a very formidable problem in devising ways to present that volume of information so that it is understandable to the users, whoever they may be. This is perhaps the most difficult and challenging problem of all.

Methodological Issues

1. *Intermedia Linkages and Tradeoffs.* Evidence of the importance of linkages and tradeoffs among the ambient qualities of the three environmental media is provided by the results of the analyses. The examples presented demonstrate the kinds and extent of the interdependencies among residuals forms and among environmental media, but the question remains: is it really worth the effort to include these linkages explicitly in the analyses? One part of the question is how difficult and costly it is to include these linkages in the analysis of alternative regional REQM strategies. The other part of this question is how costly might it be to the region if these linkages were not considered explicitly.

With respect to the first part, the Lower Delaware Valley study indicates that the marginal costs of including the air-water-solids linkages are modest—to the extent that the costs can be separated at all. There were no problems in obtaining the linkage data themselves, such as fly ash generation rates in sludge incineration, or suspended solids loads created in wet stack-gas scrubbers. The one drawback of the linked models is their size, which has implications for the costs of analysis. But all things considered, the benefit–cost ratio for linked models appears very favorable to further development and application, especially where AEQ standards are involved. The second part of the question is more difficult to answer, and no direct attempt was made in the study to obtain an answer.[17]

2. *Nonlinear Aquatic Ecosystem Models.* The potential value of using aquatic ecosystem models in analyses for REQM is substantial for those contexts in which the questions to be answered justify such models and

---

[17] There is, however, some discussion of the issue in Spofford, Russell, and Kelly, *Environmental Quality.*

the analytical resources to develop and apply them are available. Compared with classical linear dissolved oxygen models, the aquatic ecosystem models are potentially capable of providing additional information on the resulting state of the natural environment (such as algal densities and fish biomass). Also, they are alleged to provide more accurate predictions of dissolved oxygen over wider ranges of river flows and residuals discharges than the linear dissolved oxygen models. But there are substantial costs involved in incorporating them in a regional management model. For one thing, the inclusion of nonlinear ecosystem models within an optimization framework creates substantial computational problems and expense. In addition, at the current state of the art, these models do not provide accurate predictions at trophic levels above algae.

Whether or not it is worth the problems and expense of incorporating aquatic ecosystem models depends on the region being studied and the question being addressed. For the larger and more complex regions, a reasonable compromise at this time might be to use a linear, dissolved oxygen water quality model within the optimization framework. This would at least provide a relationship in the model between organic discharges and one indicator of ambient water quality. Then, using the optimal set of organic discharges from the regional model as input to an ecosystem simulation model, the implications for other water quality indicators, such as algal densities, could be investigated.

Another possibility would be to employ a linear phytoplankton model in the regional analysis. These models do exist, but they are not in as widespread use today as the nonlinear variety. In addition, certain restrictions must be placed on their use in the analysis because of the assumptions made in their development.

The feasibility of using nonlinear aquatic ecosystem models in regional REQM analyses would be enhanced by the development of efficient, large-scale, nonlinear programming algorithms that could deal with resource management problems of the type described here.

3. *Distributional Information.* The capability of providing information on the distribution of both REQM costs and ambient environmental quality is one of the most important features of the regional REQM model presented here. In most regions, the *distribution* of costs and of AEQ will be a more important issue than regional economic efficiency. Unfortunately, the provision of this additional information is not without computational problems.

Relatively little difficulty was encountered in constraining levels of ambient environmental quality and generating information on the implied costs of the constraints by political jurisdiction in the region. However, real computational difficulty arose when the attempt was made to constrain the levels of ambient environmental quality and the levels of consumer cost increases simultaneously. There are two primary reasons for this difficulty. The first involves infeasible solutions, which are commonplace when both types of constraints are imposed. This will not be discussed here.

The second computational difficulty is associated with the fact that most nonlinear programming algorithms become less and less efficient as the optimum is approached. When the regional efficiency criterion is employed, it makes sense to stop these algorithms short of an optimum because only modest cost savings, in terms of percentage of total regional costs, are involved. However, significant shifts in the distribution of costs continue to occur with virtually every step in the approach to the optimum. Thus, stopping short of the optimum makes it difficult, if not impossible, to determine when comparable results of runs for different sets of conditions have been obtained. In addition, stopping short of the optimum makes it difficult to explore the tradeoffs among the distributions of costs because these tradeoffs occur at the flattest portion of the regional cost response surface, near the optimum. Therefore, for analysis of regional REQM where specific information on distribution of both costs and ambient environmental quality is desired, resort must be made to linear programming techniques with which optimum solutions can be obtained relatively efficiently. This means the elimination of nonlinear models of both the natural world and of residuals generation and discharge as components of regional REQM optimization models.[18]

4. *Analysis with Discharge Standards.* The optimization model of the Lower Delaware Valley region had as its objective the achievement of specified AEQ standards at least cost to the region. A logical question then is, how relevant is such analysis in the real world where not only AEQ standards exist but where discharge standards or best management practices have been imposed on individual dischargers, stemming from federal and state regulatory activities? Structuring the analysis in a pro-

---

[18] This statement assumes that the nonlinearities are such that they cannot be converted into piecewise linear segments for inclusion in a linear programming model.

gramming format enables not only determining whether or not the AEQ standards can be met, given the discharge standards and best management practices, but also (A) enables determining the least-cost *additional* physical measures necessary to achieve the AEQ standards, if they are not achieved by the specified discharge standards and best management practices, and (B) enables investigating efficiently the effects on REQM costs of changes in factor prices and technology. In cases where the specified AEQ standards will clearly be achieved by the specified discharge standards and best management practices, relatively little would be gained by adopting the programming methodology. Where they will not, the additional information generated by the methodology may well be worth the additional analytical effort.

PART FOUR

Macroscale Issues and Analyses

CHAPTER 8

Macro Modeling of Residuals Discharges and Discharge Reduction Costs

As stated in chapter 2, analyses at the micro and regional levels provide useful results in and of themselves and also provide essential inputs for macrolevel analyses. However, certain critical issues being debated currently in the national arena relating to residuals management and residuals management costs can only be addressed by analyses at the macrolevel. As concern for environmental quality and restrictions on the discharge of residuals to the environment have increased, and legislation for tightening such restrictions further has been passed and is in the process of being implemented, several important questions have been raised at the national level. Can the United States afford the costs of cleaning up? What portion of the gross national product (GNP) is represented by REQM costs? How can environmental quality aspects be handled in assessing national economic growth and societal welfare?

An attempt to shed some light on these questions was the objective of the macrolevel analyses undertaken by the Quality of the Environment program. In this and the next chapter we discuss the results of those analyses with respect to the magnitude of REQM costs for the U.S. economy and the consideration of REQM impacts and costs in national income accounting.

To provide some perspective on REQM problems and costs at the national level, the Quality of the Environment program cooperated with other RFF programs—principally the population program—to develop projections into the future of residuals discharges and discharge reduction costs. This enterprise was largely designed and led by Ronald G. Ridker of RFF, and it is his report to the U.S. Commission on Population Growth and the American Future on which parts of this chapter are based.[1] An effort was made to assess the relative significance of various variables thought to be important in determining these trends—economic growth, technological changes, population increase, and intensity of residuals management effort. This research and its results have been reported in detail in the commission's report. Here we merely outline the procedures used, present some projections, and discuss the main conclusions. At the heart of the enterprise is a dynamic input–output model of various interrelationships in the U.S. economy. For purposes of this book, it is termed the macro model. More detail on the model is provided in appendix 8-A. The model is still in the process of revision and extension. The version presented here is the original one upon which the reported projections are based.

Outline of the Macro Model

A schematic diagram of the model is presented in figure 29. As the illustration reveals, the forces that "drive" the model are population size and characteristics and overall levels of economic activity or final demand. Final demand includes both the mix of goods and services and the specific characteristics of these goods and services. Final demand projections are used to develop estimates of required production levels by economic sector and year by year into the future. From these levels, plus rough estimates of the extent of materials reuse and modification of residuals that might occur over the years (under varying policy assumptions), discharges to natural environments are estimated. At the same time the model is used to calculate natural resources commodity inputs required to meet these output levels, including explicit consideration of materials reuse. We shall say very little about the latter estimates in this chapter for two reasons: (1) our main focus in this book is on REQM problems; and (2) the projections show that neither economic nor population growth in the United States is likely to be severely restrained by resources inadequacy—at least until the turn of the century.[2] All of the quantitative dis-

---

[1] U.S. Commission on Population Growth and the American Future. *Population, Resources, and the Environment,* Ronald G. Ridker, ed., Vol. III of commission research reports (Washington, D.C., U.S. Government Printing Office, 1972).

[2] It appears that the country will face some rather substantial adjustments to energy shortages and higher energy prices over the next several years as well as difficult challenges in developing new, environmentally benign, sources. Ultimately available supplies of energy in the United States are very large, however.

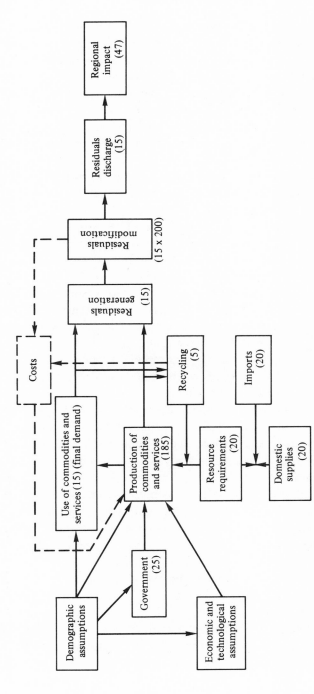

Figure 29. Macro model of interrelationships in the U.S. economy. Note: Numbers in parentheses indicate number of items or categories considered in each box. Some box titles differ from those in the original commission report. From: U.S. Commission on Population Growth and the American Future, *Population*, p. 38.

cussion pertains to the period up to approximately the year 2000. Because of rapidly mounting uncertainty as one projects further into the future, there does not seem to be much point in trying to develop detailed projections beyond this period. Some projections were run out to the year 2020, however, to see whether "disaster" lay just beyond the turn of the century; it did not. Furthermore, the projections are only for the United States. The situation in other countries, especially developing countries, appears to be quite different, but systematic research on these situations is just beginning.

The central element in the model is a dynamic input–output model of 185 economic sectors, developed at the University of Maryland by Clopper Almon and his associates. The set of consumption functions embodied in the Almon model was replaced by a set which is sensitive to demographic as well as economic characteristics of spending units and with government expenditure equations that depend on demographic characteristics as well as per capita income. In addition, several sets of coefficients were added to the input–output model—a set of resource coefficients covering some twenty commodities, and sets of residuals generation, residuals modification, and modification cost coefficients, covering some fifteen residuals, such as sulfur dioxide (SO_2), suspended solids, and biochemical oxygen demand (BOD_5). In addition, some attempts were made to allow, albeit roughly, for the possibilities of increased recycling and future substitutions of one resource for another.

In running the model, no dramatic changes in the variables were assumed. More or less steady evolution in important variables that have shown change in the past was assumed to continue. Thus, worker-hour productivity for the whole economy was assumed to continue to grow; many of the identifiable substitutions of one material for another— aluminum for steel, plastics for metals and paper, computers for semiskilled manpower, electricity for human and other forms of energy, air transport for land and water transport, and the like—were assumed to continue to work themselves out; production processes used by newer plants were taken to be spread throughout the economy; and many of the trends on consumer spending habits were assumed to continue for a reasonable period into the future. No basic changes in product mix and product specifications were assumed to occur.

Military spending was assumed to mark time for a while and later grow roughly in accord with changes in gross national product. Construction expenditures on such items as highways and education (the two biggest items) were taken to grow less rapidly than in the past; public expenditures for safety, sanitation, health, residential housing, and urban renewal were assumed to grow more rapidly; and all such rates in the model were regarded as being influenced by the rates of growth in population, in per capita income, and in the need to maintain close-to-full employment.

This set of assumptions is held constant for all runs of the model. Doing so permits focusing on the differential impact of a smaller set of variables of principal interest: population growth rates (and associated demographic characteristics such as age, labor force, and size of household), economic growth rates (insofar as they are determined independently of population growth rates), and some policy alternatives with reference to conservation and REQM.

The alternative population assumptions used in the model require a word of explanation. They center around U.S. Bureau of Census Series B and E, the former involving an average of 3.1 and the latter 2.1 children per woman.[3] The latter would lead gradually to stability of U.S. population while the former implies continued growth in U.S. population. (This assumes that there are no large-scale legal or illegal increases in immigration.)

Alternative economic growth assumptions are reflected in the model through changes in labor productivity or output per worker, which are in turn a result of differences in number of hours worked. Total output per worker-hour is assumed to continue to grow at 2.5 percent per year (about the industrial trend). Two alternative trends in working hours were specified: (1) that work hours continue to decline as they have in the past by about 0.25 percent per year; and (2) that this rate of decline increases to one percent per year. The former means a decline in the work week from approximately 40 hours to 37 hours by 2000; in the second case, 29 hours per week would be the figure for the year 2000. While this second case represents quite a substantial and perhaps unrealistically rapid increase in leisure time, it is still quite a distance from assuming an end to economic growth.

The population and economic growth assumptions lead to four basic scenarios: a high population and high economic growth case (labeled B-High or B-H); a low population and economic growth alternative (E-Low or E-L); and the two intermediate situations (B-L and E-H). Analysis of

---

[3] U.S. Bureau of the Census, *Current Population Reports,* Series P-25, no. 448 (August 1970). Labor force projections comparable with these population series were provided in unpublished form by the Bureau of the Census.

the results is begun by discussing the characteristics of the four "economies" to which these cases lead and then moves on to consider the residuals discharges projected to accompany each. (The characteristics of
scenario B-H are shown in table 41.)

For scenario B-H, two REQM policies were considered. In the first,
trends with respect to REQM policies are assumed to continue as would
have been expected as of 1967 (the base period for most of the data used).
This involves no pressure to search for and introduce dramatically fewer
residuals-generating technologies, no increased pressure to recycle, no
changes in product specifications and product mix (except as would occur for reasons other than constraints on residuals discharges), and no
taxes on the consumption of commodities that are heavy users of common
property resources. But this scenario does not mean that residuals discharges per unit of output will not change over time. It merely means that
such changes will arise as a consequence of the business executive's general interest in substituting cheaper for more expensive resources and
reducing materials and energy flows to minimize market-type costs per
unit of output. In the second REQM policy, the implications of alternative
rates of population and economic growth are assessed given more active
policies aimed at inducing a reduction in residuals discharges: more recycling, additional process change to reduce residuals generated per unit
of output, more transformation of residuals to reduce discharges of the
most harmful substances per unit of residuals generated, alterations in the
composition of demand and output and in product specifications, and so
forth. At the highly aggregated level of analysis in this chapter, it is possible to specify such policies only crudely. However, it is possible to specify
them well enough to provide some rough order-of-magnitude estimates of
the resulting discharges and REQM costs. The model demonstrates that
different REQM strategies can have vastly different costs associated with
them.

The Economy

Table 42 shows the central economic implications of the four basic scenarios, with no change in REQM policy. Significance lies in the relative
changes in GNP and other aggregates over time and among scenarios at
the same point in time, not in their absolute values.[4]

---

[4] This is because the definitions used in the model are somewhat different from
those used in official U.S. national accounts. For example, exports and imports in-

The dominant effect of shifting from Series B population projection to the Series E projection is an increase in the fraction of the population in the labor force. This change occurs for two reasons: first, because the number of persons in the labor force age range rises as a fraction of the population and, second, because additional women can be expected to enter the labor force when they have fewer children to take care of at home. The principal consequence is a higher per capita GNP and, except for the first ten to fifteen years (when because of greater female participation the labor force might actually be larger), a lower GNP under the Series E than under the Series B projections.

In terms of its principal national accounting aggregates, the composition of GNP changes only marginally over time and between runs. However, tables 41 and 42 provide a somewhat misleading picture of this situation in that they assume constant prices from 1967 on. If the same calculations had been undertaken in current dollars, one would probably observe that government spending becomes an increasing percentage of GNP, while consumption and investment as percentages of GNP fall over time. This is because service and labor prices are likely to rise more rapidly than the average price level; indeed this may happen to an even greater extent when a slower growth rate in the labor force is assumed.

The composition of both output and consumption will shift somewhat in the direction of services and away from commodities. This trend is more marked under the high productivity assumption than under the low, and under the Series E than under the Series B population assumption.

Residuals Discharges with No Change in REQM Policy

Some additional explanation of the procedure utilized to estimate residules discharges is useful at this stage. An estimate was made of the amounts of individual residuals generated by different production and consumption activities by ten-year intervals to the year 2000. (As previously explained, because the basis on which to project generation and residuals coefficients is so uncertain, no attempt was made to go beyond 2000.) Even if no change in REQM policy is assumed (from the 1967–70 base

clude only merchandise and margin items, and consumption excludes these import items plus a category called "services rendered without payment by financial intermediaries." As a consequence, GNP as determined by the model in 1967 and 1970 is respectively 4.6 percent and 4.7 percent less than officially defined GNP in those years. We have no reason to believe that this relationship will change significantly in the future.

TABLE 41. Demographic and Economic Indicators for Scenario B-High, No Change in REQM Policy

Indicators	Absolute figures					2000 compared to 1970		Percentage of total				
	1967	1970	1980	2000	2020	Ratio	Annual growth rate	1967	1970	1980	2000	2020
Population (millions)	199	205	237	321	440	1.57	1.5	100.0	100.0	100.0	100.0	100.0
Labor force (millions)	81	85	101	136	186	1.60	1.6	40.7	41.5	42.6	42.4	42.3
Households (millions)	59	62	77	106	145	1.71	1.8	29.6	30.2	32.5	33.0	33.0
GNP per capita ('67$)	3,810	3,940	5,200	8,120	12,700	2.06	2.4	—	—	—	—	—
Disposable income per capita ('58$)	2,400	2,600	3,460	5,400	8,650	2.08	2.5	—	—	—	—	—
GNP (bil. '67$)	758	807	1,232	2,608	5,571	3.23	4.0	100.0	100.0	100.0	100.0	100.0
Consumption (bil. '67$)	474	524	806	1,704	3,747	3.25	4.0	62.4	65.0	65.4	65.3	67.3
Investment (bil. '67$)	113	99	165	341	688	3.44	4.2	14.9	12.3	13.4	13.1	12.3
Government (bil. '67$)	174	186	268	579	1,170	3.11	3.9	23.0	23.1	21.8	22.2	21.0
Defense	68	62	66	97	149	1.56	1.5	9.0	7.7	5.4	3.7	2.7
Nondefense	106	124	202	482	1,021	3.89	4.6	14.0	15.4	16.4	18.5	18.3
Net exports (bil. '67$)	−3	−3	−8	−17	−33	6.22	6.3	−0.4	−0.3	−0.6	−0.6	−0.6

Total output (bil. '67$)	1,239	1,326	2,021	4,174	8,900	3.15	3.9	100.0	100.0	100.0	100.0	100.0
Primary	75	84	116	207	406	2.46	3.1	6.1	6.3	5.7	5.0	4.6
Mining	21	22	32	59	119	2.68	3.4	1.7	1.6	1.6	1.4	1.3
Construction	53	57	90	181	382	3.17	3.9	4.3	4.3	4.5	4.4	4.3
Manufacturing	567	585	880	1,776	3,689	3.04	3.8	45.8	44.1	43.5	42.6	41.4
Food	85	93	127	214	405	2.30	2.8	6.9	7.0	6.3	5.1	4.6
Paper	20	22	34	73	155	3.32	4.0	1.6	1.7	1.7	1.7	1.7
Petroleum	23	26	36	60	115	2.31	2.9	1.9	1.9	1.8	1.4	1.3
Chemicals	41	45	70	152	321	3.38	4.1	3.3	3.4	3.5	3.6	3.6
Primary metals	48	44	66	122	251	2.77	3.5	3.9	3.3	3.2	2.9	2.8
Rubber and plastics	13	14	24	54	117	3.86	4.7	1.0	1.0	1.2	1.3	1.3
Stone and clay	14	14	22	48	101	3.43	4.2	1.1	1.1	1.1	1.2	1.1
Textiles	21	24	37	69	140	2.88	3.5	1.7	1.8	1.8	1.6	1.6
Lumber and wood	12	12	20	43	92	3.58	4.3	1.0	1.0	1.0	1.0	1.0
Leather	5	4	6	14	31	3.50	4.2	0.4	0.3	0.3	0.3	0.3
Services	544	600	935	2,009	4,424	3.35	4.1	43.9	45.2	46.3	48.1	49.7
Electricity	17	19	32	72	163	3.79	4.5	1.4	1.4	1.6	1.7	1.8
Consumption purchases (bil. '67$)	474	524	806	1,704	3,747	3.25	4.0	100.0	100.0	100.0	100.0	100.0
Durables	48	52	89	207	478	3.98	4.7	10.1	9.9	11.0	12.1	12.8
Nondurables	118	128	181	333	667	2.60	3.2	24.9	24.4	22.5	19.6	17.8
Services	308	344	536	1,164	2,602	3.38	4.1	65.0	65.7	66.5	68.3	69.4

Notes: dashes = not applicable; bil. = billion.
Source: U.S. Commission on Population Growth and the American Future, *Population*, p. 40.

TABLE 42. Demographic and Economic Indicators for Four Scenarios with Alternative Population and Economic Assumptions, No Change in REQM Policy

| Indicator | 1970 | Absolute figures | | | | | | | | Percentage reduction from B-H in 2000 | | | Percentage reduction from B-H in 2020 | | |
| | | 2000 | | | | 2020 | | | | E-H | B-L | E-L | E-H | B-L | E-L |
		B-H	E-H	B-L	E-L	B-H	E-H	B-L	E-L						
Population (millions)	205	321	266	321	266	440	299	440	299	17.1	0	17.1	32.0	0	32.0
Labor force (millions)	85	136	127	136	127	186	146	186	146	6.6	0	6.6	21.5	0	21.5
Households (millions)	62	106	101	106	101	145	113	145	113	4.7	0	4.7	22.1	0	22.1
GNP per capita ('67$)	3,940	8,120	9,100	6,450	7,220	12,700	14,600	8,630	9,950	(12.0)[a]	20.6	11.2	(15.5)[a]	31.8	21.4
Disposable income per capita ('58$)	2,600	5,400	6,020	4,240	4,720	8,650	9,850	5,800	6,560	(11.5)[a]	21.4	12.6	(13.8)[a]	32.9	24.2
GNP (bil. '67$)	806	2,607	2,419	2,072	1,920	5,572	4,373	3,797	2,973	7.2	20.6	26.4	21.5	31.8	46.6
Consumption (bil. '67$)	524	1,704	1,577	1,339	1,237	3,747	2,899	2,514	1,930	7.5	21.4	27.4	22.6	32.9	48.5
Investment (bil. '67$)	99	341	309	278	252	688	551	492	400	9.4	18.5	26.1	19.9	28.5	41.9
Government (bil. '67$)	186	579	548	468	442	1,170	948	813	659	5.4	19.2	23.7	19.0	30.5	43.7
Defense	62	97	102	88	92	149	158	128	133	(5.2)[a]	9.3	5.2	(6.0)[a]	14.1	10.7
Nondefense	124	482	446	380	350	1,021	790	685	526	7.5	21.2	27.4	22.6	32.9	48.5
Net exports (bil. '67$)	-3	-17	-15	-13	-11	-33	-25	-22	-16	(11.8)[a]	(23.5)[a]	(35.3)[a]	(24.2)[a]	(33.3)[a]	(51.5)[a]

Total output (bil. '67$)	1,326	4,174	3,843	3,334	3,064	8,900	6,933	6,124	4,747	7.9	20.1	26.6	22.1	31.2	46.7
Primary	84	207	192	174	161	406	318	297	231	7.2	15.9	22.2	21.7	26.8	43.1
Mining	22	59	57	49	47	119	100	87	72	3.4	16.9	20.3	16.0	26.9	39.5
Construction	57	181	164	148	133	382	296	274	209	9.4	18.2	26.5	22.5	28.3	45.3
Manufacturing	585	1,776	1,628	1,437	1,316	3,689	2,877	2,587	2,012	8.3	19.1	25.9	22.0	29.9	45.4
Food	93	214	198	184	170	405	312	306	234	7.5	14.0	20.6	23.0	24.4	42.2
Paper	22	73	66	58	53	155	119	107	82	9.6	20.5	27.4	23.2	31.0	47.1
Petroleum	26	60	62	51	52	115	105	87	76	(3.3)[a]	15.0	13.3	8.7	24.3	33.9
Chemicals	45	152	142	120	112	321	254	218	173	6.6	21.1	26.3	20.9	32.1	46.1
Primary metals	44	122	111	100	90	251	197	180	140	9.0	18.0	26.2	21.5	28.3	44.0
Rubber and plastics	14	54	48	43	38	117	88	80	60	11.1	20.4	29.6	24.8	31.6	48.7
Stone and clay	14	48	43	39	35	101	79	74	56	10.4	18.7	27.1	21.8	26.7	44.6
Textiles	25	69	60	52	46	140	102	89	65	13.0	24.6	33.3	27.1	36.4	53.6
Lumber and wood	12	43	39	34	31	92	71	63	49	9.3	20.9	27.9	22.8	31.5	46.7
Leather	4	14	10	11	7	31	20	20	12	28.6	21.4	50.0	35.5	35.5	61.3
Services	600	2,009	1,858	1,575	1,454	4,424	3,442	2,966	2,295	7.5	21.6	27.6	22.2	32.9	48.1
Electricity	19	72	66	58	53	163	127	114	88	8.3	19.4	26.4	22.1	30.1	46.0
Consumption purchases (bil. '67$)	524	1,704	1,577	1,339	1,237	3,747	2,899	2,514	1,930	7.5	21.4	27.4	22.6	32.9	48.5
Durables	52	207	188	159	144	478	362	314	236	9.2	23.2	30.4	24.3	34.3	50.6
Nondurables	128	333	313	278	261	667	525	482	377	6.0	16.5	21.6	21.3	27.7	43.5
Services	344	1,164	1,076	902	832	2,602	2,012	1,718	1,317	7.6	22.5	28.5	22.7	34.0	49.4

Notes: B-H = high population and high economic growth alternative; E-L = low population and growth alternative; B-L and E-H = intermediate alternatives; bil. = billion.
Source: U.S. Commission on Population Growth and the American Future, *Population*, p. 41.
[a] Percentage increase from B-H.

259

period), the coefficients utilized to estimate these levels still change over time. This is because of changes in production processes, end-product specifications, and input mixes that as far as could be foreseen would be coming along anyway.

For example, in the pulp and paper industry, there has been a major shift in the pulping processes used in the last twenty-five years. Of the approximately 10 million tons of wood pulp produced in 1945 in the United States, about 18, 23, and 44 percent were produced by mechanical (groundwood), sulfite, and sulfate processes, respectively. The percentages of the approximately 41 million tons produced in 1969 were about 10, 9.5, and 67.5, respectively. Sulfate pulping results in the generation, per ton of unbleached pulp, of about half as much biological oxygen demand (BOD_5), about the same amount of suspended solids, about one-sixth as much sulfur dioxide (SO_2), and twice as many particulates as magnefite pulping with 80 percent recovery of spent pulping liquor. (Higher recoveries, economically justifiable, are common.) In addition, sulfate pulping results in the generation of the odorous reduced sulfur compounds. Generation of color residuals is much higher for sulfate pulping than for sulfite pulping. (The comparison for integrated pulp and paper mills producing the *same* brightness *product* [paper, not pulp] would be in the same direction, although not as pronounced, because of the relatively greater increases in residuals generation in sulfate bleaching.) Given the ability of the kraft process to pulp any type of wood species, it is likely that much of the increment in pulp production (from wood) in the future will be by this process.

At the same time, there has been a trend in final product specifications that has operated to increase residuals generation per unit of paper or paper product produced. The principal trend in product specifications has been toward higher brightness products, thereby requiring more bleaching, which is a major source of residuals generation in paper production. For example, using the sulfate (kraft) process, shifting to the production of unbleached tissue (napkin, towel) paper from the current brightness specifications would reduce generation of BOD_5 by about 80 percent, using present typical bleaching sequences.

Insofar as possible, tendencies of this kind were identified in other industries. Where there was found to be a major discrepancy between generation coefficients for the most efficient plants, in items of materials use and residuals discharges, and for the average of the industry, it was assumed that today's most efficient plant would become the norm by the year 2000. But changes in technology beyond today's "best practice" were not introduced. This determined the levels of residuals discharges without an "active" REQM policy. The procedure followed to obtain discharges under an active REQM policy is presented in the next section of this chapter.

Attention should be called to the fact that the projections presented deal solely with discharges and say nothing about degrees of deterioration in ambient environmental quality. To do this, models of natural systems are required, such as those discussed in preceding chapters. These are not applicable when discharges are treated in large geographical aggregates such as a nation. Moreover, when REQM policies were instituted in the analysis, they were on a nationally uniform basis. These are severe restrictions but unavoidable in a macro analysis of this kind.

Table 43 presents residuals discharges for the nation as a whole for 1970 and projections for 2000, assuming no change in REQM policy. Three principal conclusions emerge from these data. First, if the intensity of residuals modification is not increased, even a substantial slowdown in the growth rates of both population and economic activities from the rates prevailing in recent years will not be enough to keep the levels of residuals discharges in the year 2000 below current levels.

But future rates of growth in these residuals will be lower than they appear to have been in recent years. They are lower than assumed rates of growth in GNP; and for some residuals, they are even lower than growth rates for GNP per capita and for population. These results arise principally from two factors. One, composition of output in the year 2000 involves slightly less generation of residuals than does the 1970 composition, mainly (but not entirely) because of a small, but definite, shift away from goods toward services. Two, while a few individual residuals generation and discharge coefficients are expected to increase over time (for the kinds of reasons noted in the above pulp and paper illustration), more are expected to decrease as best present practice becomes pervasive.[5]

While one may be fairly confident about the trends for the residuals listed in table 43, an important qualification must be entered for residuals not listed in this table. A number of material residuals, such as heavy metals, synthetic organics, pesticides, and several sources, such as agriculture, mining, urban storm runoff, silviculture, are not considered in the calculations. Furthermore, it may be that more advanced industries

---

[5] Increase in energy costs may also lead to lower residuals generation, as already has occurred in some cases.

TABLE 43. Residuals Discharged Under Four Scenarios, No Change in REQM Policy

Residuals	1970	Quantity discharged (billion pounds) 2000				Ratio 2000/1970				Percent reduction from B-H in 2000		
		B-H	E-H	B-L	E-L	B-H	E-H	B-L	E-L	E-H	B-L	E-L
Gaseous												
Particulates	41	49	45	41	38	1.20	1.10	1.00	0.93	8.2	16.3	22.4
Hydrocarbons	89	207	196	181	171	2.33	2.20	2.03	1.92	5.3	12.6	17.4
Oxides of sulfur	85	132	122	108	100	1.55	1.44	1.27	1.18	7.6	18.2	24.2
Carbon monoxide	231	493	470	489	466	2.13	2.03	2.12	2.02	4.7	0.8	5.5
Oxides of nitrogen	30	58	55	54	51	1.93	1.83	1.80	1.70	5.2	6.9	12.1
Liquid												
Waste water[a]	188	323	307	296	283	1.72	1.63	1.57	1.51	4.9	8.4	12.4
COD[b]	202	462	416	398	359	2.29	2.06	1.97	1.78	10.0	13.8	22.3
BOD$_5$[c]	68	134	121	115	103	1.97	1.78	1.69	1.51	9.7	14.2	23.1
Suspended solids	804	1,590	1,440	1,350	1,230	1.98	1.79	1.68	1.53	9.4	15.0	22.8
Dissolved solids	137	294	265	251	225	2.15	1.93	1.83	1.64	9.9	14.6	23.5
Nitrogen	23	43	38	37	33	1.87	1.65	1.61	1.43	11.6	13.9	23.3
Phosphorus	8	16	15	14	12	2.00	1.88	1.75	1.50	6.2	12.5	25.0
Thermal-to-water[d]	5.6	3.55	3.30	2.90	2.65	0.63	0.59	0.52	0.39	7.0	18.3	25.4
Solid[e]	321	863	718	860	715	2.69	2.24	2.68	2.23	16.8	3.5	17.1

Note: The figures in this table differ from those in the original population commission report in that they contain estimates of thermal discharges to water bodies. These were not contained in the original figures. The estimates for thermal discharges were made by Chris Sandberg of RFF.

Source: U.S. Commission on Population Growth and the American Future, *Population,* p. 46.

[a] Trillion gallons.

[b] COD = chemical oxygen demand.

[c] BOD$_5$ = biochemical oxygen demand (5 day).

[d] 10^{15} Btus. The conservative assumption was made that all power plants of 500 Mw or greater capacity would install cooling towers under the present REQM policy. Although not likely to be universally true, the increasing constraint on water availability would tend in this direction.

[e] Trillion pounds; includes sludge generated in modifying liquid residuals.

will generate new kinds of residuals in the future at a rate that invalidates these results, that adverse effects of some presently generated and discharged residuals will be discovered, or both.[6] Unfortunately, from our analysis we have no way of knowing.

Residuals discharges seem to be related somewhat more to growth in the economy than to growth in the population, although there are marked differences among different kinds of residuals. Solid residuals of the types included and carbon monoxide, for example, appear more related to population, whereas sulfur oxides, suspended solids, and phosphorus are more closely associated with the level of economic activity. The principal explanation is that a slowdown in population growth induces some offsetting increase in economic growth, whereas an independent slowdown in economic growth involves no comparable offset.

Residuals Discharges and REQM Costs with an Active REQM Policy

In the present context, an active REQM policy means a policy that induces reductions in residuals generation, discharge coefficients, or both.[7] This can be done in a number of ways. In fact, much of this book is devoted to discussing and assessing these alternatives. Here it is simply assumed that discharges are limited to certain specified levels, with reductions being achieved principally through residuals modification processes; then the effect that this has on aggregate discharge levels and costs of reduction is observed.

The effluent standards used for this demonstration are roughly the 1973 water and the 1975 air discharge standards recommended by the Environmental Protection Agency (EPA) before the passage of the 1972 Water Quality Act Amendments. These effluent standards were accepted for this exercise only after it was ascertained that it is technologically feasible (or soon will be) to meet them without assuming any dramatic technological breakthroughs. For a number of sectors and residuals, methods of reducing generation and discharge coefficients to meet the standards are already used on a sufficiently large scale that one can have reasonable confidence in the cost estimates. An important case in point is waste water treatment where the EPA standards call for "secondary treatment." As

---

[6] The historic record has notable examples of the "time lag" in recognizing the harmful effects, that is, DDT, PCBs.

[7] But, as we will see later, this by no means exhausts the list of policy options—nor is it necessarily always the most attractive.

shown in the last chapter, this is a well-established procedure (at least for municipal sewage plants) that removes 85 to 90 percent of the BOD_5 and suspended solids from municipal liquid residuals. In other cases—for example, oxides of nitrogen reduction and sulfur removed from fuel—where physical measures that would meet the standards have only recently begun to be used commercially, it was ascertained that at least one option was available, the most feasible one accepted, and cost estimates were developed accordingly. It is quite possible that between now and the year 2000 considerably more stringent effluent standards will be imposed and achieved, and costs will be accordingly higher—perhaps much higher (see the following section).[8] But the standards used here are sufficient to reduce discharges considerably below present levels, even a year-2000 levels of population, production, and consumption. For the projections, the assumption is made that measures needed to reach the standard will be introduced gradually between now and the year 2000.

Before proceeding to a discussion of the results, it should be noted that the projections procedure was not explicitly able to account for the interdependencies among residuals streams. As the reader will infer from our earlier discussion, this is a significant omission. Because of this and a number of other inadequacies and uncertainties, the final interpretation of results must regard the quantitative outcomes as merely order-of-magnitude estimates.

Table 44 presents the results of the application of the active REQM policy for a representative set of residuals. The rows labeled A, provided for background, are the levels of residuals that would be generated using 1970 technology so far as residuals generation coefficients are concerned. The rows labeled B indicate actual discharges for 1970, and estimated discharges for the year 2000, assuming no change in REQM policy from the base period. A comparison of rows A and B indicates the extent to which changes in generation and discharge coefficients likely to come along anyway will reduce discharges. The C rows indicate the level of discharges likely in 2000, assuming the active REQM policy outlined above. It can be seen that such a policy would reduce the discharges of all included residuals significantly below current levels. This is true except for dissolved solids where standards are not very stringent, partly because available modification technologies are very expensive and partly because

---

[8] Under the assumption of no changes in basic production process, product mix, product specifications, and so forth. Such changes could substantially reduce costs to achieve high levels of discharge reduction.

TABLE 44. Residuals Generated and Discharged under Alternative Assumptions: A, B, and C[a]

(billion lbs.)

Residuals	1970	Quantity discharged			
		Year 2000			
		B-H	E-H	B-L	E-L
Gaseous					
Particulates					
A	54.0	175.0	160.0	154.0	134.0
B	41.0	49.0	45.3	41.4	38.3
C	—	5.6	5.2	4.7	4.4
Hydrocarbons					
A	96.0	270.0	254.0	233.0	219.0
B	89.0	207.0	196.0	181.0	171.0
C	—	31.4	29.4	27.4	25.6
Oxides of nitrogen					
A	30.0	80.0	76.0	73.0	68.0
B	30.0	58.0	55.0	54.0	51.0
C	—	18.0	17.0	17.0	16.0
Liquid					
Biological oxygen demand					
A	72.0	186.0	167.0	160.0	144.0
B	68.0	134.0	121.0	115.0	103.0
C	—	17.0	15.0	14.0	13.0
Suspended solids					
A	846.0	2,000.0	1,810.0	1,700.0	1,530.0
B	804.0	1,590.0	1,440.0	1,350.0	1,230.0
C	—	99.0	90.0	84.0	76.0
Dissolved solids					
A	137.0	363.0	329.0	305.0	276.0
B	137.0	295.0	265.0	251.0	225.0
C	—	285.0	256.0	243.0	217.0
Thermal (to water)[b]					
A	5.6	3.55	3.30	2.90	2.65
B	5.6	2.95	2.75	2.40	2.20
C	—	0.0	0.0	0.0	0.0

Notes: The figures in this table differ from those in the original population commission report in that they contain estimates of thermal discharges to water bodies. These were not contained in the original figures. The estimates for thermal discharges were made by Chris Sandberg of RFF. Dashes = not applicable.

Source: U.S. Commission on Population Growth and the American Future, *Population*, p. 48.

[a] A = Residuals generated assuming 1970 technology, that is, no changes in residuals generation coefficients (technological changes resulting in substitutions among inputs—e.g., plastics for metals— are included, however). B = Residuals discharged in 1970, and in 2000 assuming changes in residuals generation coefficients and efficiency of modification likely to come along even without an active REQM policy. C = Residuals discharged in 2000 assuming use of production and modification processes induced by an active REQM policy. Figures for B-L and E-L in row C are approximations, interpolated from the B-H and E-H estimates calculated in detail.

[b] 10^{15} Btus. The conservative assumption was made that all power plants of 500 Mw or greater capacity would install cooling towers under the present REQM policy. Although not likely to be universally true, the increasing constraint on water availability would tend in this direction.

some modification technologies involve the addition of chemicals, thereby increasing dissolved solids.

Table 45 indicates the annualized costs associated with residuals modification under the active REQM policy. It does not include a number of cost items that a broader REQM policy would involve, for example, additional water storage facilities, separation of storm and sanitary sewers, correction of damages caused by strip mining, and control of nonpoint sources of residuals—that together would no doubt entail large sums. It does, however, contain those items that are conventionally included in official estimates of residuals modification costs. In 1970, such costs amounted to about $9 billion (1967 dollars); by 2000, the figure could amount to more than $50 billion, thus growing at a faster rate than GNP. However, the percentage of GNP represented by these costs would be less in 2000 than in 1980.

The period of most rapid rise appears to be the next decade. Thereafter, because we may have caught up somewhat with past neglect and because sufficient time will have elapsed to find less costly ways to reduce residuals discharges (such as substitute materials, products, processes with less residuals generation) the increase in costs could be much less rapid. The same need to catch up in order to achieve the assumed standards suggests that in 1980 the savings involved if the lower rather than the higher population projections were to occur would be negligible. By the year 2000, however, these savings would amount to about 10 percent.

Thus, although not huge over the time period we are considering, the difference between rapid continued population growth and a tendency toward stabilization would nevertheless be significant. If one were to consider the additional problems associated with congestion in cities, roadways, and recreation areas, the differences might be considerably greater. In any case, a tendency toward population stabilization would be helpful in coping with REQM problems. Various changes—availability and attitudes toward the use of birth control devices—seem, in fact, to be carrying the United States in that direction in any case. The nation is now at, or slightly below, a net reproduction rate of one. If this persists, U.S. population will stabilize at a level about one-third higher than at present (again discounting the possibility of large-scale increases in immigration).

The costs of improving ambient environmental quality will be large. As discussed below, they will be considerably larger than the costs shown in table 45 because of various omissions from the projections shown

TABLE 45. Annualized Cost of an Active REQM Policy under Two Population Assumptions and High Economic Growth (absolute numbers in billions of $1967)

Environmental media	1970	REQM costs 1980		2000		2000 compared to 1970 Ratio		Annual growth rates		Percent savings from
		B-H	E-H	B-H	E-H	B-H	E-H	B-H	E-H	B-H
Air	0.45	11.7	11.7	14.9	14.1	32.0	31.2	12.3	12.1	5.5
Water	3.26	9.7	9.5	18.2	16.8	5.58	5.17	5.7	5.4	7.6
Land	5.18	10.6	10.5	18.8	16.9	3.62	3.25	4.4	4.0	10.2
Total	8.89	32.0	31.7	51.9	47.8	—	—	5.9	5.6	7.9
Percent of GNP	1.10	2.60	2.64	1.99	1.97	—	—	—	—	—

Notes: The figures in this table differ from those in the original population commission report in that they contain estimates of the costs of reducing thermal discharges to water bodies. These were not contained in the original figures. The estimates of these costs were made by Chris Sandberg of RFF. The figures do not include estimates of costs of reducing discharges of residuals from urban storm runoff, agriculture, silviculture, and mining. An interest rate of 6.5 percent was used in deriving the annualized costs. Dashes = not applicable.

Source: U.S. Commission on Population Growth and the American Future, *Population*, p. 49.

therein. But from such expenditures will be derived the benefits of cleaner air and cleaner water and a less disrupted landscape. Other research in the Quality of the Environment program analyzed some of these benefits, such as the benefits to human health from improving ambient air quality.[9] However, because all of the benefits associated with reducing the discharge of all of the gaseous, liquid, and thermal residuals and associated with better handling of solid residuals cannot as yet be defined in monetary terms, it was not possible to derive aggregate national benefits from REQM comparable to aggregate national costs.

Overview of REQM Costs

Fundamentally, the costs of improving ambient environmental quality represent real resources of capital, labor, and raw materials devoted to cleanup—resources that otherwise have been available for building homes, educating children, constructing mass transit systems, building and maintaining vacation facilities, and for the host of other goods and services bought by individual citizens and state and local governments. The costs are not simply figures of interest to economists, accountants, and engineers. Mention of sources of costs not included in the figures cited in the previous section is merited.

With respect to water quality, there are several large sources of liquid residuals that pose potentially major and costly REQM problems. Liquid residuals from agricultural operations, such as nutrients and pesticides, as well as residuals contained in runoff from feedlots and large poultry farms, may have important adverse effects on water quality. Coal mining activities can be major sources of acid drainage; mining activities in general, including sand and gravel operations, can be major sources of suspended solids. Oil spills and disposal of dredging spoil can have major adverse effects on water quality. The REQM costs associated with all of these are poorly defined.

Urban storm runoff can be another major source of adverse effects on water quality. In many major cities, storm and sanitary sewers are combined and feed into sewage treatment plants. During rainstorms, the sudden surge in the volume of water exceeds treatment plant capacity, and raw sewage is then spilled into the receiving water bodies. Separating storm from sanitary sewers would require massive capital outlays in most

[9] See Lester B. Lave and Eugene P. Seskin, *Air Pollution and Human Health* (Baltimore, Johns Hopkins University Press for Resources for the Future, 1977).

metropolitan areas. Even after separation, storm water, if not modified, would deposit substantial amounts of residuals (for example, dissolved and suspended solids and metallic ions) into water bodies. Building lagoons to hold the storm water, which would then be fed gradually into treatment plants, has been suggested as one alternative. In any event, the appropriate technologies and the costs of dealing with this problem are highly uncertain.

With respect to air quality, the costs of reducing residuals discharges from automobile travel are still uncertain. Automobile manufacturers are currently planning to use exhaust recirculation and various catalytic devices in combination with the standard internal combustion engine. While there is disagreement on this approach, it is very likely that it will prove to be the most expensive alternative, in terms of the initial cost of the automobile, reduced gasoline mileage, higher operating costs, and lower performance.[10] More radical approaches, relying on different kinds of engines, would be much less costly in the long run, according to many experts.[11]

Even with the stringent controls on residuals discharges from automobiles, there will be a number of major American cities that will not be able, in the late 1970s, to meet ambient air quality standards without imposing major restrictions on travel. There would be at least some offsetting benefits in terms of reduced congestion, reduced traffic control expenditures, and reduced need for expensive parking facilities. Estimation of the net costs of the multiple reallocations involved is complex, yielding substantial uncertainty in those estimates.

With respect to the costs of managing solid residuals, perhaps the largest uncertainties in estimating such costs relate to disposal of sludge, animal manure, and mining overburden (tailings, and solid residuals from ore processing). For example, as restrictions on discharges of liquid residuals are tightened, modification of the liquid residuals to meet those restrictions results in increasing quantities of sludge for disposal. The problem is particularly acute in metropolitan areas, where land costs make the traditional sludge-drying beds prohibitively expensive. Incineration

[10] Recent evidence indicates that the catalytic devices themselves will result in the generation of other harmful residuals, a perfect example of the need to consider explicitly the interrelationships among types and forms of residuals in developing REQM strategies.

[11] For example, see Robert U. Ayres and Richard P. McKenna, *Alternatives to the Internal Combustion Engine* (Baltimore, Johns Hopkins University Press for Resources for the Future, 1972).

of sludge requires additional inputs—particularly energy—and results in the generation of gaseous residuals, as well as a "final" solid residual. As yet, neither alternative methods of sludge disposal and their costs, nor the costs of traditional methods of disposal are well defined. Similarly, the costs of disposing of the solid residuals from mining and ore processing are uncertain.

Because of the various uncertainties, any estimate of total REQM costs for the nation to meet any specified set of discharge standards, ambient environmental quality standards, or both, should be specified for a wide range. By the early 1980s, annual REQM costs might well be in the range of $50 to $100 billion. The $50-billion level would represent about 10 percent of the growth in per capita national income that would otherwise have been available to increase living standards and conditions in other directions. It is quite possible that REQM costs could absorb up to one quarter of this if all ambient environmental quality and residuals discharge goals are achieved by the deadlines specified in current legislation.[12]

Concluding Comments

Methods for estimating future REQM costs are still crude, and the future necessarily contains many elements of uncertainty. Governmental agencies, particularly the Environmental Protection Agency and the Council on Environmental Quality, are continuing to refine such estimates. The range in estimated costs is substantial, even for a given set of residuals discharge standards. But a further difficulty lies in the dynamic nature of those standards (even if all other relevant variables remain constant). Both litigation and legislation are in process that may well result in significantly different targets for certain residuals and certain residuals-generating activities. Toxic materials and strip mining are but two of many examples.

It would seem, therefore, that the importance of all this is how these numbers are affected by the major variables among them and the proportion of GNP represented by REQM costs. At the same time, various estimates suggest that the benefits from reducing the discharge of residuals

[12] For a discussion of some possible short-run cost implications of recent federal pollution control legislation, see Allen V. Kneese and Charles L. Schultze, *Pollution, Prices, and Public Policy* (Washington, D.C., The Brookings Institution, 1975).

will be substantial. The results of the main studies do suggest, however, that if the nation approaches its REQM problems with reasonably efficient programs and permits some time for adjustments to be made, it will be possible to reduce residuals discharges very substantially below current levels while at the same time enjoying economic growth at only a slightly diminished rate over the remainder of this century. This would be true whether high or low population growth occurs. Low population growth does yield a considerable saving in REQM costs in the longer run (beyond 1980 to 1985) and appears to be highly desirable on a number of other grounds. REQM costs will be large; their size emphasizes the importance of paying close attention to cost effectiveness in the design of REQM policies and programs.

Appendix 8-A
A More Precise Discussion of the Macro Model

This appendix describes the model developed by Resources for the Future to project economic activity, resource requirements, and residuals loadings. In the body of this chapter we introduced the general framework and assumptions of this model; here we outline the structural relationships involved and indicate the way in which they were employed for the purposes of projection and policy analysis.

The core of this model is the 185-sector University of Maryland Interindustry Forecasting Model developed and maintained by Clopper Almon and the staff of the Interindustry Forecasting Project at the university. This core was modified and added to in a number of ways to be discussed in more detail below. These modifications involved extension of the time horizon from 1980 to the year 2000, modifications to make the model more sensitive to demographic changes, the introduction of a mechanism to permit changes in technology on a sector-by-sector basis, and the internalization of the public component of final demand, treated as an exogenous input in the original model.

In addition, a number of subsidiary models were developed. The complete package, which, for convenience, we shall refer to as the RFF macro model, is presented schematically in figure 8-A-1. In contrast to the more

AUTHORS' NOTE: This appendix is based on material prepared by Ronald G. Ridker.

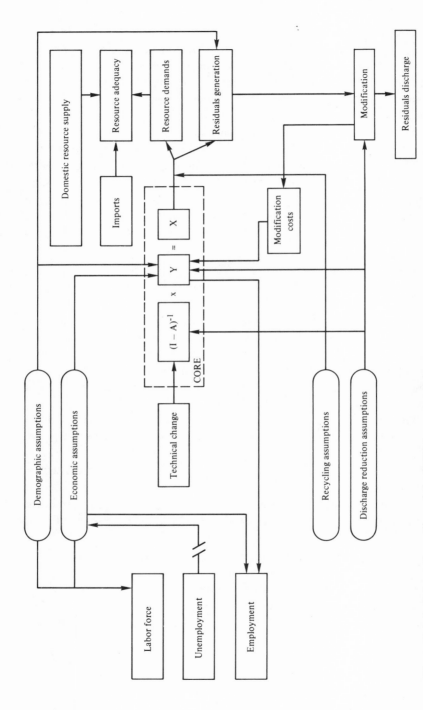

Figure 8-A-1. Schematic outline of the RFF macro model

general conceptual layout presented in the chapter, this figure is organized on a computational basis, the arrows representing flows of information provided by one stage or submodel to another.

The Core Model

The core model may be described in static terms using standard input–output equations

$$AX + Y = X$$

where

X = column vector (185 × 1) of total outputs
Y = column vector (185 × 1) of final demands
A = 185-order matrix of input–output coefficients a_{ij}

where a_{ij} = million dollars of purchases by industry j from industry i required per million dollars of total output produced in industry j.

Given a forecast of final demand Y and the satisfaction of the usual input–output theorems such that $[I - A] \neq 0$, then

$$X = (I - A)^{-1}Y$$

The model is linked to time in a number of important ways. First, it is recursive, requiring some prior year projections to obtain future year projections. For this reason, the model moves forward a year at a time, a projection to the year 2000 requiring 34 iterations. Second, technical change is handled explicitly by allowing the technical coefficients a_{ij} to change over time.[1] Accordingly, the linear relationship that links total output to final demand is itself a function of time. Third, each equation that predicts sectoral personal consumption expenditures as a function of several independent variables contains two shift coefficients that alter the relative importance of these variables over time. The degree to which these coefficients operate at the sectoral level depends on several projected demographic characteristics of families.

The exogenous information required by the core model defines the specific scenario under consideration. In general, this information consists of projected values for future years for each of the following: disposable

[1] A discussion of the technique of doing this is found in chapter 12 of U.S. Commission on Population Growth and the American Future, *Population*.

income per capita, population, school age population, labor force, and number of households and their distribution by age of head and size. All but the first of these exogenous inputs to the model are defined by the particular population series projection selected for consideration. Once such a series has been selected, various time streams of disposable income per capita may be chosen to reflect alternative levels of individual well-being including, besides the monetary measure, preferences for leisure. The choice of an appropriate time stream for per capita income is also dependent upon assumptions about growth in labor productivity and acceptable unemployment rates. Final choice of a time stream for this variable, as well as for a number of public expenditure variables, must be made on a trial and error basis, the aim being to obtain a satisfactory unemployment rate.

As figure 8-A-1 indicates, this exogenous information influences the model through its effect on sectoral final demands. For example, disposable income per capita, population size, and household characteristics determine the total level and relative distribution of personal consumption expenditures. The final demand vector Y, in figure 8-A-1, is the summation over a number of vectors, each representing a different category of final demand.

The forecasting method for the sectoral values of the final demand vectors is specific to each category of final demand—personal consumption expenditures, gross private domestic investment, and government purchases of goods and services. For reasons of space, in this appendix we will discuss only the projection of the personal consumption component. It is the largest component, accounting for more than 60 percent of GNP, and it provides some of the key linkages between changes in demographic characteristics and the resources and environmental consequences of these changes. The projection technique for other components of final demand can be found in the population commission report cited earlier.

Personal Consumption Expenditures

The Maryland input–output model identifies some 126 items of personal consumption and provides an equation for each. These equations take the general form

$$c_{it} = \hat{b}_{i0} + \hat{b}_{i1}Y_t + \hat{b}_{i2}p_{it} + \hat{b}_{i3}t + \hat{b}_{i4}\Delta Y_t \qquad (1)$$

$$C_{it} = c_{it}P_t \tag{2}$$

where the subscripts i and t represent sector number and time respectively, and

c = per capita consumption
C = total consumption
Y = disposable per capita income
$\Delta Y_t = Y_t - Y_{t-1}$
p = price index for the sector relative to the overall price index for all sectors combined
P = population
t = time in years

$\left.\begin{array}{l}\hat{b}_0 \\ \hat{b}_2 \\ \hat{b}_3 \\ \hat{b}_4\end{array}\right\}$ = parameters estimated by regressions, and

b_1 = parameter estimated algebraically

To minimize changes in the original model, RFF retained the general form of these equations but "conditioned" them to reflect changing family demographic characteristics.

The parameters b_{i1} in equation (1) are derived from the total expenditure elasticities n_i^0, as follows

$$b_{i1} = n_{i1}\left(\frac{c_{it}}{Y_t}\right) \qquad t = 1961 \tag{3}$$

Cross-section consumption surveys were employed to provide information relevant to the pattern and not the levels of personal consumption expenditures. A time series regression covering the period 1958 through 1969 was then performed for each input–output consumption expenditure category. The dependent variable in each case was that portion of consumption not explained by income. Thus,

$$z_{it} = \hat{b}_{i0} + \hat{b}_{i2}p_{it} + \hat{b}_{i3}t + \hat{b}_{i4}\Delta Y_t \tag{4}$$

subject to

$$|\hat{b}_{i4}| \leq \hat{b}_{i1} \qquad \text{if} \qquad \hat{b}_{i4} \leq 0$$

where

$$z_{it} = c_{it} - b_{i1}Y_t$$

As the equations stand, the only demographic linkage is provided by the total population variable, P_t, a strict multiplier to per capita consumption in equation (2). A certain amount of demographic influence could possibly enter through the taste change parameters $\hat{b}_{i3}$; however, one would suspect that the short period over which the regressions were performed would impart very little, if any, of this influence to the regression estimates. Thus, at this point, the estimated forecasting equations for per capita consumption are, for all practical purposes, "demographic-neutral."

The required demographic linkage is provided by a reweighting procedure on the cross-section expenditure survey data. For each population projection for the year 2000, a two-way distribution of families by age of family head and size was developed and used to generate new total expenditure elasticities for the 126 consumption forecasting equations.

However, it is not these new elasticities, but the components that determine them, that are employed as "shift parameters" in a modified version of equation (1). Graphs such as that in figure 8-A-2 can be constructed from the cross-section expenditure data for any particular item of consumption i. The two relationships that are plotted reflect how much of good i is consumed, on a per capita basis, at different levels of total expenditure. The line marked "base" is the relationship determined from the cross-section information using the actual family characteristics existing in the base year. The other line marked "scenario K," is the relationship obtained under the reweighting technique already mentioned.

The total expenditure elasticity, n_i^0, employed in equation (3), is the ratio of the marginal to average propensities to spend on good i, computed using the base period weights. In figure 8-A-2, this elasticity may be represented as

$$n_i^0 = \frac{c_0 - c_0^*}{\bar{\epsilon}} \bigg/ \frac{c_0}{\bar{\epsilon}}, \text{ or}$$

$$n_i^0 = \frac{c_0 - c_0^*}{c_0}$$

where $\bar{\epsilon}$ is the average total expenditure on all goods. In a like manner, the total expenditure elasticity characterizing scenario K may be represented by

$$n_i^k = \frac{c_k - c_k^*}{c_k}$$

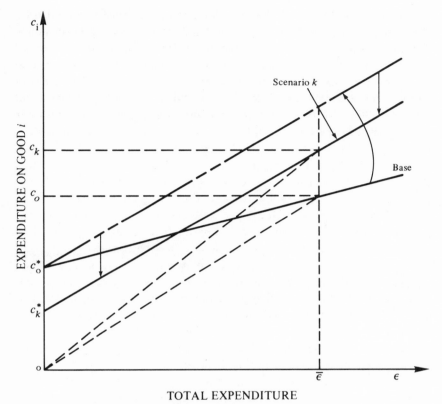

Figure 8-A-2. Propensity to spend on good i

To capture the character of the shift portrayed in figure 8-A-2, two adjustments, or shift parameters, have been introduced. The first involves the rotation of the base line about the fixed intercept point, c_0^*, or a change in the manner in which individuals spend on good i out of increased (or decreased) income (total expenditures). If scenario K is representative of the year 2000, and the rotation is assumed to begin in 1970 and apply equally over the 31-year period, then for any year t^*, the new income parameter is equal to

$$b_{i1} + t^*\delta_i \qquad t^* = 0 \text{ in } 1969$$

where

$$\delta_i = \frac{n_i^k - n_i^0}{31}\left(\frac{c_{it}}{Y_t}\right) \qquad t = 1961$$

The second adjustment to equation (1) involves a shift downward (or upward) in the base line, once rotated, in figure 8-A-2. This shift in intercept from c_0^* to c_k^* is interpreted, irrespective of the direction or magnitude of the first type of adjustment, as a change in the relative importance of non-income factors in determining the level of consumption of good i. Like the first shifter, this adjustment will be applied to equation (1) in 31 equal increments. It will be entered as a percentage change in cross-section intercept to the non-income determinants of per capita consumption. Thus, for any year t^*, this second shift parameter is equal to

$$1 + t^*\phi_i \qquad t^* = 0 \text{ in } 1969$$

where

$$\phi_i = \frac{(c_k^* - c_0^*)/c_0^*}{31}$$

Because of the length of the projection period, the private savings rate s was entered in an explicit manner as a control on the total personal consumption expenditures over the forecast period. Equations (1) and (2) may now be restated to include all of the above refinements

$$C_{it}^* = v_t C_{it}$$

where

$$C_{it} = \{(b_{i1} + t^*\delta_i)Y_t + (1 + t^*\phi_i)[\hat{b}_{i0} + \hat{b}_{i2}p_{it} + \hat{b}_{i3}t + \hat{b}_{i4}\Delta Y_t]\}P_t$$

and

$$v_t = \frac{(1 - s)Y_t P_t}{\displaystyle\sum_{i=1}^{126} C_{it}} \qquad t = 0 \text{ in } 1966 \qquad t^* = 0 \text{ in } 1969$$

Residuals

Future residuals discharges into the environment originate from two basic sources in the RFF macro model. This can be seen in figure 8-A-1 by the two arrows that enter the box marked "residuals generation." One type results from the production of goods and services to satisfy the final demands of the economy. This type of residuals, hereafter termed residuals from production, is tied directly to sectoral total outputs in the model. The other type of residuals comes from the consumption of final goods and is thus termed residuals from final consumption. This type is related to the magnitude of variables best considered under the demographic assumptions of the model. These include the projections of population size and its distribution between urban and rural locations, number of automobiles and miles traveled, and number of sewered dwelling units.

Fifteen categories, or types, of residuals are estimated in the model to be discharged to three receiving media: air, water, and land. The gross (unmodified) residual of type k generated in year t from source i, G_{it}^k, is forecast in the model by equations of the form

$$G_{it}^k = a_{it}^k B_{it}^k,$$

where a_{it}^k is a time-dependent parameter for source i and residual type k, and B_{it}^k is the source, or residual-generating base, defined as

$$B_{it}^k = \begin{cases} x_{it} \text{ or total output, for residuals from production} \\ \text{Population, automobile mileage, and the like, for residuals} \\ \text{from final consumption} \end{cases}$$

The total gross residual of type k generated in year t is

$$G_t^k = \sum_i a_{it}^k B_{it}^k$$

If modification measures for type k residuals achieve an overall modification efficiency of b_t^k in year t, then the net residual discharge, N_t^k of this residual in year t is

$$N_t^k = (1 - b_t^k) \sum_i a_{it}^k B_{it}^k$$

The next-to-last equation indicates that the submodel built to forecast residuals discharges has gross generation coefficients a_i^k for each residual, that are not only specific to source, but time dependent as well. The modification efficiencies, on the other hand, although time dependent, are not specific to the source. Estimates of the costs of achieving the modifications specified were made by the application of cost coefficients, that is, dollars per thousand pounds of BOD_5, particulates, and the like, removed, to the number of units of discharge reduction (units removed).

CHAPTER 9

REQM Considerations in National Income Accounting

We have seen that the costs of reducing the discharge of residuals, while still quite uncertain, will be large. They are substantial enough to raise concerns about how such costs, and the changing service flows from the common property resources they are intended to protect, should be reflected in our national income accounting. At present, such aggregative economic measures as national production and labor productivity are affected differently, depending upon whether REQM outlays are made in the public or the private sector. Accordingly, some adjustments are indicated if only on the grounds of consistency. These macroeconomic issues have received both conceptual and quantitative attention in recent years in the Quality of the Environment program (and in the economics profession generally). This chapter describes the results of our analyses of the national income accounting issue.

Modification of National Accounts to Reflect Residuals

Should gross national product (GNP) and net national product (NNP) be modified to account for ambient environmental quality deterioration and the costs of reducing that deterioration? The question of whether the aggregate output accounts should be modified to reflect the growing generation, handling, and discharge of residuals can be interpreted in two ways. First, should the official definitions of the national income and product accounts be changed, and, second, should auxiliary modified series be presented along with the official series based on unchanged definitions? Generally it appears that the official series should be continued on the basis of the present definitions, both because of the desirability of avoiding breaks in the series and also because the potential advantages and the significance of changes that might be made are not yet completely clear. The discussion here applies, then, to the second interpretation. Whenever we speak of the desirability of modifying GNP or NNP, we refer not to the official series but to modified auxiliary series that could be associated with the existing series in various ways. Experience with such new series might later suggest the desirability of a change in the official definitions.

Gross national product is designed to measure the production of final goods and services in the economy as a whole. The final consumers of these goods and services are taken to be individuals and households (consumers in the usual sense), government, exports, and nonprofit institutions. The assumption is made that these particular economic agents do not usually use inputs to provide intermediate services (such as, for example, a transportation company would), but rather that they "use up" the utility embodied in the goods and services that the economy produces. This is a working assumption to which there are numerous exceptions.

At any given time there exists a list of goods and services that is officially defined as "final."[1] This set of goods and services is exchanged in markets, and therefore there is a corresponding set of market-exchange-determined prices attached to them in some base year as well as in the current period.[2] As time passes, new types of goods and services are often "wedged in" to help keep the list more nearly complete. To calculate price-corrected or real GNP, the changing numbers of physical units of the final goods and services produced are multiplied by the unchanged base-period prices—currently those of 1958. The system of national accounts is of the double entry type. In *current* prices the total of GNP calculated from the product side must balance GNP calculated as the sum of values added of all activities contributing to GNP (value added is the value of sales minus cost of purchased inputs). This is not true of price-corrected, or *real*, GNP, however, because no deflator has been devised with which the value-added side could be price corrected.[3] Accordingly,

---

[1] As discussed later, governmental services do not appear as a list of final products. These services are evaluated on the basis of inputs purchased.

[2] In fact, some of these transactions are virtual or imputed. For example, the value of owner-occupied housing is estimated by imputation.

[3] Deflated gross national product by industry is calculated by deflating industry outputs and purchases separately and subtracting. See U.S. Department of Commerce, Office of Business Economics, *Concepts and Methods of National Income Statistics* (Washington, D.C., GPO, 1954) p. 26. (NTIS no. PB-194 900 [OBE-SUP 70-02]).

since this discussion deals with real GNP, reference here will be to the product-side calculation.

If all salient goods and services were exchanged in markets; if the degree of competition in these markets did not change; if the programs of government and nonprofit institutions did not change in ways that alter substantially the welfare they produce relative to the final goods and services that they absorb; if population stayed constant; and if the distribution of income did not change; then alterations in real NNP (GNP minus capital consumption allowances) could be taken to be a good indicator of changes in the economic welfare of the population.

This is an imposing string of assumptions, however, none of which corresponds exactly to reality. To the extent that they are violated, NNP diminishes in usefulness as a welfare measure. In fact, the gap between reality and this set of assumptions is large and significant in some cases. (To take a pertinent instance, the existence of common property resources is neglected.) Accordingly, the usefulness of NNP as a welfare measure is severely limited.

Of course the system of national accounts—which includes much more than total production, for example, a large set of industrial sector breakdowns—has been designed to serve a number of purposes. Even the aggregates serve multiple purposes. They are intended to provide information for short-run economic stabilization policies and programs, and they are intended to furnish an estimate of the total production of goods and services that the society has available to meet alternative goals. To be fair, one must recognize that the designers of the accounts thought that at best they would serve to provide only a rough indicator of one dimension of welfare of a society.

But to segregate discussion of the accounts entirely from broader questions of welfare, as some students and practitioners of accounting would do, is a serious mistake. Whether it was the original intent or not, NNP or GNP are now often explicitly or implicitly viewed as indexes of welfare, and changes in them, as changes in welfare. Furthermore, to understand what the accounts do measure, it is important to recognize explicitly that there are large flows of services and disservices, valued by people, that do not enter into market exchange and which therefore are not in the list of final goods and services. Unless care is taken to recognize and identify these flows, NNP may become grossly misleading about what is happening even to production of potentially marketable goods and services in the economy. For instance, should there be a large-scale transfer from purchases of services (for example, house painting, grass cutting, construction, household services) to self-provision of these services by households, the NNP would tend to fall, whereas it need not be true that production would have declined. The reason that NNP would tend to fall is that the labor going into these self-provided services is not among the defined final products, so that working time shifted toward them disappears from the account. The reason that such services provided to themselves by households are not in the official list is that the accountants have found them too difficult and costly to identify and evaluate. However, there could be reason to reconsider this position in view of current data collection and handling techniques and in view of the many attempts to define such matters as "value of time" in commuting and leisure activities.

It is illuminating to regard the objections of the environmentalists to the present accounts as revolving around the question of what is or is not in the list of final products.[4] When the environmentalists argue that GNP overstates growth (and hence welfare), they are implicitly incorporating in the list of final products many entirely real service flows which, however, do not enter into market exchange and accordingly are not in the official list of outputs. Moreover, environmentalists implicitly believe that the net effect of bringing omitted service flows into the accounts would be to reduce the growth in *real* product. Such a list would probably include, for example, the life support, aesthetic, and convenience services of high quality air, high quality water, and spacious surroundings—all of which, as we have already seen, are in some of their aspects common property resources not entering into private exchange. The only way a change in these service flows could now influence the aggregate measures of national output is if their changed quality or quantity made the production of items which in fact are included in the list easier or more difficult, for example, having to treat intake water or air in order to produce a good. In reality, at least up to now, such feedbacks to the national accounts from altered quality of the common property resources are probably trivial compared with alteration of service flows from these resources direct to final consumers, that is, damages from lower air, water, and land quality. The latter are nowhere reflected as such in the list of final products, although they may affect some items that are. It is the deterioration in the quality of the environmental services not included in the official list that mainly concerns the environmentalist.

---

[4] This is not to imply that only environmentalists are concerned with this issue.

The exclusion of the services of air, water, and land from the list is not primarily the result of disagreement that the services provided by nature are a major factor in true human welfare. Rather it reflects primarily a judgment on the part of the income accountant that obtaining acceptable estimates for these values would be too difficult and costly to be worth the effort. Clearly, however, any reduction in the service flows of common property resources that is viewed as a loss in real product by consumers means that NNP overstates any increase in final product as compared with the total flow from the truly welfare-relevant and much larger list of final goods and services. In the extreme case it is actually possible that the "true" service flow could decrease while NNP rises. Some environmentalists believe that this is happening now. Whether such is the case or not is an empirical question that cannot be answered by theoretical arguments.

That burdens on the service flows from common property resources tend to rise with increasing production unless effective, collectively imposed REQM measures are undertaken is obvious from observation and from considering the implications of mass balance. As we have indicated, conservation of mass implies that all material resources used as inputs to the extractive, productive, and consumptive activities of the economy must appear as residuals. These have in some manner to be returned to the environment except, of course, for changes in the inventory of mass. If the use of materials should rise faster than production of final goods and services, residuals generation must too. There are counteracting trends affecting materials use in the economy. But the fact that, as lower quality ores are used, greater quantities of unwanted material must be processed and more energy used to obtain a given quantity of wanted material implies a tendency for residuals to rise faster than final production of goods and services embodying materials. Also, energy usage recently has been rising more rapidly than real NNP, and so long as it is obtained primarily from conversion of fossil fuels, this implies a rapidly rising flow of residual materials as well as residual energy itself. In fact, it appears that net energy output per unit of energy input is decreasing, that is, deeper wells to extract crude petroleum, where effects of depth are greater than effects of improved technology. Other sources of nonlinearities can be readily identified. Indeed, sometimes discontinuities or thresholds are encountered, as when a water body becomes anaerobic and its ecological functioning changes dramatically for the worse, insofar as providing services such as recreation and fishing are concerned.

Most national accountants would probably agree that if it were practical to extend the list of final goods and services to include service flows from the natural environment, this should be done. But the difficulties of doing it are truly imposing.[5] Consequently, it is often concluded that the best we can do is to supplement the real NNP with physical, chemical, or biological indicators of the state of the environment. This has been advocated and seems to be a very good idea. However, the methodology is still in a rather primitive state of development. We discuss one such possible extension in a following section.

But are there any less ambitious adjustments that could be made to the monetary accounts themselves? One possibility would be to deduct consumer defensive expenditures from the NNP. If environmental service flows remained constant, then defensive expenditures made voluntarily by consumers would be on the same footing as any other consumer expenditures, being carried to the point where utility gained is equated with alternative cost in utility lost. It would make no difference if environmental service flows are included in the list of final products so far as indication of welfare changes over time is concerned.

If environmental service flows change, however, then it is clear that a list of final products that omits either these flows or the defensive consumer expenditures may give an incorrect indication of welfare change over time. If defensive expenditures were simply deducted from the present NNP, the necessary implicit assumption would be—if welfare change is to be correctly indicated—that these defensive expenditures exactly offset the decline in value of the environmental services that "ought to" but do not now affect the NNP. Even if this strict assumption is not accepted, it probably would still be of general interest to try to estimate consumer defensive expenditures.

Defensive expenditures by industry are already appropriately treated from the exclusion point of view because they never appear in real NNP. We will not develop the explanation of this behavior of the accounts here because it is quite similar to that developed in some detail for REQM expenditures below.

---

[5] However, at the time of this writing, a substantial effort was underway to provide a conceptual, and the beginnings of a quantitative, basis for including environmental service flows in the accounts. See Henry M. Peskin, "A National Accounting Framework for Environmental Assets," *Journal of Environmental Economics* vol. 2 (1976) pp. 255–262.

Costs of Meeting Ambient Environmental Quality Standards

Up to this point we have been discussing common property environmental resources with regard to the national accounts as though their use were completely unrestricted. In the United States this would have been generally true until the last decade, but as we have seen, public policy is now evolving rapidly to regulate the use of these resources for residuals disposal. This policy has led to the development of ambient environmental quality standards and subsequently to the development of residuals discharge standards on a local, state, and national basis. If these standards become effective, they will give rise to large expenditures for the control of residuals generation and discharge. The time pattern these expenditures will follow is uncertain, but, in the previous chapter, we speculated that they will "hump" in the next five to ten years during a clean-up phase. Thereafter, they may decline slightly for a time, following which they may tend to rise nonlinearly with increasing output. The question is, How should these expenditures be handled in the NNP?[6]

Before trying to answer, it is necessary to understand how such expenditures are handled under present practice. Perhaps surprisingly, they are handled differently depending on whether they are incurred by consumers and government on the one hand or by industry on the other. In the following discussion we will neglect expenditures for residuals management *directly* by consumers, because we think they will be small (with one major exception discussed later), and in any case no different principles are involved for them.

The differential effect on the accounts of industry and government expenditures for residuals management can be illustrated by a simple example, shown in table 46. Assume an economy in which only two commodities, haircuts and bread, are produced in the base period (the citizens will be nude but well clipped). The list of final products will, therefore, consist of haircuts and bread. Assume further that the production of haircuts generates no significant amount of residuals but that the production of bread does. Suppose also that barbers can be diverted to reduce the discharge of residuals if that is desired (the bread can be produced with fewer residuals if more labor is used).

---

[6] In the following discussion the major emphasis is on NNP because GNP includes depreciation of capital assets, which cannot in any sense be regarded as contributing to the nation's economic welfare.

TABLE 46. Effect of Expenditures for Residuals on Industry and Government Accounts

Period	Haircuts, bread	Quantity	Unit price (dollars)	Net national product (dollars)
Period 0	H	100	10	1,000
	B	100	10	1,000
				2,000
Period 1a	H	50	10	500
(as it would be without limitation	B	150	10	1,500
on residuals discharges)				2,000
Period 1b	H	25	10	250
(as it would be if industry had to	B	150	10	1,500
reduce residuals discharges by				1,750
diverting 25 barbers)				
Period 1c	H	25	10	250
(as it would be if government hired	B	150	10	1,500
the 25 diverted barbers and set	Gᵃ	25	10	250
them to reducing residuals				2,000
discharges)				

ᵃ G = government.

In the base period (period 0) there is a standard for the discharge of residuals, but the output of bread is just low enough to avoid violating it. In period 1, a change in family composition causes a shift in demand from haircuts to bread, with a resulting increase in residuals generation. If there had been no limitation on discharge of residuals, the situation would be that labeled 1a; that is, $500 worth of productive services would have been diverted from the production of haircuts to production of bread, and residuals discharges would have increased.

We have assumed, however, that there is a limitation on residuals discharges, but it is not being met in situation 1a. If it is met by a diversion *within the industry* of barbers to reduce residuals discharges, NNP will register a decline as compared with period 0. This is shown by 1b. In contrast, if the government hires these same men to limit the discharge of residuals by the same amount, NNP will show no decline, as in period 1c. The reason for these results is that there is nothing in the list of final products corresponding to residuals management activities, so such activities cannot be reflected in NNP evaluated at base year prices.[7] However, since government is in effect regarded as a final consumer, its expenditure for the barbers (converted into residuals modifiers) is included in NNP.

Observations on Treating the Costs of Meeting Environmental Quality Standards

In a situation in which governments establish effective environmental quality standards that must be continuously met—residuals discharge standards, ambient environmental quality standards, or both—NNP treatment of industry outlays would seem to indicate the direction of welfare change more appropriately than the present treatment of similar government outlays. The reason is that the net outlays made for residuals management can be viewed as simply being necessary to maintain the service flows naturally provided by the common property resources. In that sense they could be regarded as expenditures necessary to maintain the unproduced capital stock. Failure to treat them this way could result in an anomalous situation in which a progressively larger share of production would have to be devoted simply to maintenance of ambient environmental quality with NNP all the time continuing to rise. It would be hard to claim that this rise could in any way be regarded as indicating increased welfare, even in the limited sense of increased availability of marketable goods and services contributing to consumer satisfaction. If this view were accepted, the appropriate procedure would be to continue to treat industry outlays for residuals management as at present but to change the procedures in such a way that government outlays for residuals management would be treated in the same way as industry outlays. This would require identification of government expenditures for residuals management, which should not be very difficult, and their subtraction from the present NNP for presentation as an auxiliary series.[8]

The one major exception to the view that consumer expenditures to control residuals will be small is the cost of reducing residuals discharges from automobiles. Because the cost of the control devices is initially borne by the manufacturer and not by the consumer directly, the effect on the

[7] Industry costs incurred to counter internal environmental quality problems (for example, occupational health and safety) would have the same effect. But because they are internal to the plant, presumably no one would question them.

[8] Such expenditures should include administrative expenditures as well as capital and operations and maintenance expenditures on facilities.

accounts is similar to that of other industrial residuals management costs. The approach that has recently been adopted in developing the official series is essentially to add items called "control devices" to the list of final products. If this were not done, the price deflator would tend in the direction of reduced real production of automobiles.

This approach is inconsistent with the view concerning appropriate handling of industrial outlays expressed above. To treat the automobile consumer symmetrically, would be to regard him as producing a service for himself, the production of which service generates residuals. He is then required in the interest of maintaining the service flow from common property assets to incur a cost. To add such costs to NNP over time would have the same anomalous results as already described in the case of industry. In this case, as with activities to reduce residuals from industrial activities, the accounts should be permitted to function as they normally would.

As we have already indicated, our view on these matters appears to be contrary to the views of some experts in national income accounting. Their position apparently rests on two major considerations. The first reflects the special problem of dealing with a catch-up phase, such as we are now experiencing in connection with the reduction of residuals discharges. When effective standards inducing higher than prevailing ambient environmental quality are first set, some of the expenditures made will result in actual improvements in ambient environmental quality over its current state, and the benefits of the improvement would presumably exceed the costs of achieving the improvement. Thus, if residuals reduction expenditures are not included in the accounts, the anomalous situation would arise in which the population actually experiences an improvement in welfare while the associated influence on NNP is downward. Some have proposed that the list of final products be expanded to include industrial outlays for residuals management just to avoid this situation. For reasons already explained, this approach appears to be in conflict with that appropriate from the longer term welfare-indicator point of view. It would seem preferable not to add the outlays but to prorate them over a longer period of time, especially to earlier periods when NNP tended to be overstated as an indicator of welfare.

The second consideration is that excluding production directed toward residuals management would distort labor productivity series which, at the aggregative level, are obtained by dividing NNP by worker-hours worked. The proper point of view on this is highly contingent on what is regarded as being measured by labor productivity. If it is taken to be the output per worker-hour net of the output needed to maintain the service flows of all assets—private and common—then it is wholly appropriate that productivity should tend to fall if a larger proportion of total effort has to be used for meeting environmental quality standards.

Conclusions about National Accounts

The national product and income accounts are not directed very closely to the single objective of measuring changes in social welfare. Furthermore, considerable uncertainty surrounds some of the possible changes that have been discussed—both as to implications and practicality. Accordingly, it would not appear wise to change the official definitions at this time. However, series should be prepared that reflect, at least in part, the growth of activities and expenditures that are apparently needed simply to maintain naturally provided ambient environmental quality and that would permit various types of adjustments to be made by individual players of the accounting game as each deems appropriate.

To this end, we suggest the regular preparation and publication of the following series:

> industrial expenditures for residuals management
> government expenditures for residuals management
> consumer expenditures for residuals management
> consumer, industry, and government defensive expenditures

None of these series is prepared currently, and the preparation of each offers considerable difficulty, but these can probably be overcome. In addition, it probably would be desirable to publish auxiliary series supplementing the current official series. The following are some of the possibilities, with all series assumed to be price corrected:

1. GNP—including all residuals management in the list of final products
2. NNP_1—GNP minus net depreciation of private assets
3. NNP_2—GNP minus net depreciation of private assets and minus the nonindustry cost of residuals management and defensive expenditures (and in principle all other costs that may be induced by the growth process itself, for example, congestion costs)

It would be mistaken to think, however, that any such adjustments can come close to indicating changes in "true" welfare as far as flows of environmental services are concerned. The essential absent ingredient is a valuation for the environmental services themselves and, on capital account, an expression for decreases or increases in the value of the corresponding natural assets.

Apart from this general caveat, the empirical computation of some of the series is likely to be "slippery." This is especially true of industrial outlays made for residuals management to meet environmental quality standards. This is immediately evident merely by considering the possible responses of industry to increases in the stringency of the standards.[9] True, industry may make some outlays, both capital and operating, which would be clearly designated as residuals management outlays. These might include equipment costs and operating expenses for extracting unwanted materials from stack gases or liquid residuals streams, for example. There are other responses, however. A material may be substituted that is less harmful when discharged as a gaseous, liquid, or solid residual. A basic process may be changed that cuts discharge of residuals and simultaneously produces marketable outputs or increases productivity. Product output specifications may be changed that result in lower generation of residuals per unit of product output. Residuals management processes may cause a reduction in consumption of the article in question by increasing cost and price. There is little hope for estimating the latter types of responses on any comprehensive basis.[10]

Defensive expenditures will also be very difficult to estimate accurately. How much washing is induced by deterioration in environmental quality, and how much is habit or the pleasure obtained from a warm bath on a cold day? Thus, the whole problem cannot be solved. Even though the suggested adjustments are partial and have limited accuracy, they ought to be made, for they may permit some sharpening of conclusions on what

[9] See chapter 4 for a detailed discussion of these responses. The discussion there also illustrates the difficulty of determining what proportion of the expenditure should be allocated to residuals management and what proportion to improving productivity, and so forth.

[10] In some cases, incremental costs attributable to meeting environmental standards can be estimated relatively easily—for example, the additional cost of aesthetically pleasing poles for electricity transmission lines compared with the cost of standard poles. But even in this case the environmental standard may stimulate a response which involves no additional, or even lower, costs.

has happened to real product and hence permit a better estimate of impact on welfare.

In the next section we turn to a different kind of possible adjustment of the national accounting system. This one does not pertain to the monetary summaries but rather to some of the physical flows which accompany market exchange, but not all of which are priced by it.

A Nonmonetary Extension of the Accounts

As we have pointed out several times, a basic cause of environmental quality problems is the increasing flow of residual materials and energy from the various economic sectors to the environmental common property resources. This suggests the possibility of supplementing the monetary accounting system with a materials and energy flow accounting system, based on the concepts of materials and energy flow and balance. The discussion herein deals only with flows of materials, but is equally applicable to flows of energy.

Such a procedure, if successfully developed and implemented, should substantially increase the usefulness of the national accounting system as a component of an environmental quality information system for policy makers and analysts. Specifically, the modification in the accounts would help illuminate the important relationship between the size and composition of industrial, transportation, commercial, agricultural, mining, and forestry activities and residuals generation, as well as the relationship between alternative industrial, agricultural, mining, and forestry processes and overall residuals generation. Moreover, since the modification could rely to a large extent on existing information sources, the information cost should be manageable.

National accounting systems describe the flows of goods and services between sectors, defined by a variety of criteria, such as geography (for example, foreign versus domestic account), type of ownership (for example, public versus private enterprises), type of product produced (for example, manufacturing versus agriculture), and so forth. Generally, as we have seen, the goods and service flows are measured in market-determined value units; indeed, the existence of a defined market is an important criterion for inclusion.

The basic procedure for expanding the accounts to include materials flows would be: (1) to expand the list of commodity and service flows to include all physical nonmarket inputs and outputs (regardless of whether they are conventionally considered as residuals); and (2) to emphasize process distinctions as the basic criterion of sector definition.

The Quality of the Environment program engaged Henry Peskin to elaborate such a procedure and to do an illustrative application for the Norwegian economy. Norway was chosen because it combines a relatively simple economy with comparatively good data. In this brief discussion it is not appropriate to go into the procedure Peskin developed in detail— this has been done in a separate monograph[11]—but a summary of the case application to the Norwegian pulp and paper industry will indicate the nature of the analysis and results.

The sectors selected for the illustrative application were the Norwegian sulfite pulp industry (international standard industrial classification [ISIC] 2721) and the sulfate or kraft pulp industry (ISIC 2722). These industries were chosen for two reasons. First, a comparison of these two accounts after they have been expanded to include the flows of nonmarket inputs and outputs served well to illustrate the gain in information resulting from defining sectors according to process rather than by product. In the Norwegian accounts, a full accounting of even the market-transacted inputs and outputs is made only for the aggregation of these two sectors: chemical pulp. Although sulfite pulp and kraft pulp can be close substitutes for certain uses, they differ substantially in their use and generation of nonmarket inputs and outputs.

The second reason for illustrating the accounting concepts with data for the chemical pulp industry is that available information permitted detailed estimation of the use and generation of nonmarket commodities without new data collection. This was possible because a Resources for the Future study of the pulp and paper industry (see chapter 4) yielded flow charts which could be used to identify nonmarket physical outputs associated with the processes. These flow charts were used in the following manner. For the base year (1969), each of the sixteen establishment survey questionnaires of the Norwegian Central Statistical Office for the sulfite industry and the six questionnaires for the sulfate industry were examined to determine: (1) what process was employed (by scanning the list of reported inputs), (2) how much pulp was produced, (3) what proportion of the pulp was bleached, and (4) what proportion of the pulp was dried. With some modification, which need not concern us here, the

[11] Henry M. Peskin, *National Accounting and the Environment,* Artikler 50 (Oslo, Statistisk Sentralbyrå, 1972).

appropriate flow charts were then applied to each establishment in order to estimate total use of air and water and total generation of a long list of residuals. As many as three flow charts would have to be applied if the establishment in question produced pulp of which some or all was bleached or dried (one chart for pulping, one for bleaching, and one for drying).

Table 47 shows the expanded accounts for the two industrial sectors after the appropriate aggregations were made. It reveals substantial differences between the sulfite and sulfate sectors with respect to their relative use and generation of nonmarket inputs and outputs. Also, the composition of these nonmarket items differs substantially from the composition implied for the aggregate of these two sectors.

For example, the large generation of dissolved solids for the chemical pulp industry (776 kilograms per ton of pulp) is dominated by the dissolved solids generated by sulfite processing where more than a metric ton of these dissolved solids per ton of pulp is produced. (The amount of course depends on the extent of recovery of materials in the pulping liquor.) On the other hand, the sulfate process generates per ton of pulp far more dissolved chlorine (negligible for sulfite), reflecting the fact that sulfate pulp requires relatively more bleaching to achieve the same brightness.[12] The information provided by these detailed expanded accounts shows, among other things, that the amounts and types of residuals generated in a nation's economic activities can differ greatly depending on the processes selected, even without the installation of residuals management equipment.

Large differences in the amounts and types of residuals generated also occur because of product mix and product specifications, as discussed in chapter 3 and illustrated specifically in chapter 4. Of course, product and process are directly related, but simply indicating the differences in residuals generated among processes does not make explicit the effect of product mix/product specifications. Thus, it would be useful to develop expanded accounts to reflect these factors as well. This would provide society with a clear indication of the residuals–environmental quality implications of alternative life styles.

Materials and energy accounts would be valuable tools for understanding the sources of residuals generation for national or regional economies, how the sources are related to economic structure, and what tendencies

---

[12] It should be noted, however, that sulfite processes have limitations in terms of the species that can be pulped.

TABLE 47. Comparison of Selected Estimates of Nonmarket Inputs and Outputs, Norwegian Chemical Pulp Industry (ISIC 2721 and 2722)

	Sulfite[a] (2721)		Sulfate (2722)		Total (2721 + 2722)	
	Total (metric tons)	Per ton of bleached pulp	Total (metric tons)	Per ton of bleached pulp	Total (metric tons)	Per ton of bleached pulp
Input						
Water	116.8×10^6	190.7 t	20.0×10^6	100.8 t	136.8×10^6	168.0 t
Air	0.901×10^6	1.5 t	1.32×10^6	6.6 t	2.22×10^6	2.74 t
Output[b]						
Liquid						
Water	110.9×10^6	181.0 t	17.9×10^6	90.2 t	128.8×10^6	159.0 t
Total dissolved solids	627.0×10^3	1,023.3 kg	3.04×10^3	15.3 kg	630.0×10^3	776.0 kg
Organic	624.2×10^3	1,019.8 kg	0.51×10^3	2.6 kg	624.7×10^3	770.0 kg
Inorganic	2.8×10^3	4.5 kg	2.53×10^3	12.7 kg	5.3×10^3	6.5 kg
Fiber	1.8×10^3	2.9 kg	0.11×10^3	0.56 kg	1.9×10^3	2.4 kg
Carbohydrates plus chlorine combined	0.0	0.0	2.62×10^3	13.2 kg	2.6×10^3	3.2 kg
Gaseous						
Cl₂	238.0	0.388 kg	23.0	0.116 kg	261.0	0.32 kg
Residuals sulfides	0.005×10^3	0.008 kg	1.940×10^3	9.76 kg	1.945×10^3	2.40 kg
SO₂ total	40.3×10^3	65.7 kg	1.48×10^3	7.46 kg	41.8×10^3	51.5 kg
from processing	32.3×10^3	52.6 kg	0.35×10^3	1.77 kg	32.6×10^3	40.2 kg
from fuel oil[c]	8.03×10^3	13.1 kg	1.13×10^3	5.69 kg	9.16×10^3	11.3 kg
Solid[d]	660.0×10^3	1.08×10^3 kg	24.7×10^3	124.0 kg	687.0×10^3	847.0 kg

Total pulp production 1969:
Sulfate pulp 198,700 metric tons est. dry weight
Sulfite pulp 612,600 metric tons est. dry weight

Note: t = metric ton.
Source: Peskin, National Accounting, p. 52.
[a] These figures reflect the distribution of Norwegian establishments among the various sulfite processes.
[b] Assumes no residuals modification equipment or other residuals processing.
[c] Assumes S content of 2 percent.
[d] Includes substantial water content. Dry value for sulfite is approximately 30,600 metric tons and for sulfate approximately 3,000 metric tons.

they may exhibit in the future under alternative assumptions about technological and product changes and REQM policies. They would also provide a firmer and more internally consistent data base for projection models like that discussed in the previous chapter. Work directed toward development of such expanded accounting systems is proceeding at several institutions in the United States and elsewhere.

PART FIVE

Conclusion

CHAPTER 10

Reprise

We have described a research program that addresses many problems in the complex field of residuals–environmental quality management (REQM) at the micro, regional, and macro levels. In the process a large number of concepts, methods, and results have been discussed. A concise review of the most basic of these results follows.

About Residuals

1. All human activities result in the generation of some material or energy residuals, or both. There is no process producing goods or services that transforms 100 percent of the inputs into desired outputs, nor is there any consumption process that generates no residuals. The residuals from production and consumption activities are typically discharged to the common property environmental media.

2. The external diseconomies resulting from the discharge of residuals into the three environmental media are normally quantitatively negligible in a low-density, low-production setting, but become progressively more important as density of development and level of output (affluence) rise, in the face of essentially finite reservoirs of residuals assimilation capacity.

3. These external diseconomies cannot be efficiently handled by considering and managing the three environmental media—air, water, and land—separately. Conservation of mass dictates that the weight of a residuals stream once generated cannot be reduced by modification. In fact, mass is increased, as well as is the quantity of energy residuals, because the modification process itself requires inputs. Modification processes can only: (a) change a residual into another type of the same form; (b) change the form, for example, from liquid to solid; (c) change the time, or location (or both) of discharge; and (d) any combination of (a), (b), and (c). Further, production process changes that reduce dis-

299

charges to one medium, even though they can reduce the total mass of residuals generated, may increase residuals flow to another medium. Accordingly, liquid, solid, gaseous, and energy residuals streams are generally interdependent, and the cost of modifying one will depend to a greater or lesser degree on the constraints imposed upon discharges of the others.

4. Final demand, the mix of goods and services desired by society—including the spatial distribution of activities, product mix, specifications of product outputs and services, in sum, the "lifestyle" of a society—is the driving force of the REQM problem, in terms of both residuals generation and REQM costs.

5. Recycling of materials and energy can reduce the total materials and energy flowing through a society only up to the level of recycling where the marginal material costs of recycling equal the marginal material costs of using virgin materials, including the externalities in both cases. The more material residuals are dispersed, the more inputs of materials and energy are required to accomplish any given level of recycling.

6. Both residuals generation by individual activities and the capacity of the environment to assimilate residuals vary significantly from day to day, seasonally, and from year to year. Such variation is superimposed on the dynamic context of REQM in terms of changing spatial patterns and levels of activities, technology, and social tastes.

Major Conclusions

With the above background of general propositions, we now state some major conclusions from the research program described in the preceding chapters.

1. REQM is affected by a wide range of decisions made by the individual economic activity—industrial plants, mines, agricultural and silvicultural operations, households, commercial and institutional operations, recreational operations, and residuals-handling activities of governmental agencies and private enterprises. These decisions involve choices among:

A. types of material (quality characteristics) and energy inputs
B. process technology or design of the production function, as in the design of a hospital

C. extent of materials and energy recovery and of by-product production

D. the product mix and the quality specifications for the outputs of goods, services, or both

E. the overall level and time pattern of the production of goods, services, and energy

These factors are of course interrelated. To produce a product with certain characteristics often limits the type of raw material and production process used. Furthermore, the product output specifications limit the amounts of secondary materials that can be used as inputs; for example, steel scrap in steel manufacture, paper residuals in paper manufacture. Nevertheless, the above list does usefully characterize some of the important economic and technological choices made by individual activities, choices which affect REQM.

2. REQM at the level of the individual activity is affected by many decisions and factors *external* to the activity, over which the activity has no control. Examples include: prices of energy, water, and fuel; prices of other raw material inputs, for example, chemicals, ores, secondary materials; freight rates; tariffs; and input restrictions. Among others, these factors influence decisions on the levels of materials and energy recovery and by-product production that would take place in the absence of any constraints on residuals discharges to the environment.

3. Decisions at the individual activity level may affect residuals generation: on-site, within the region, outside the region, or various combinations of all three. Thus, an on-site action such as a decision by the management of an office building to shift from bleached paper towels to unbleached paper towels—all other product characteristics remaining the same—will not change solid residuals generation in the office building, but will change residuals generation in the production of paper towels, which may well occur outside the region in which the office building is located. Similarly, many ways in which use of electrical energy can be reduced within a residence will not change residuals generation at the residence, but will reduce residuals generation within the regional power system, outside of it, or both.

4. As noted in the introductory section of this chapter, REQM takes place in a dynamic context that has continually changing factor prices, technology, product mix, product specifications, and social tastes. As the

industry studies described in chapter 4 demonstrate, failure to understand the effects of these changing factors can lead to substantially different results from a particular REQM strategy than anticipated. For example, changes in certain exogenous variables—such as the price of iron ore relative to steel scrap—can have substantial effects on residuals discharged from a steel mill, even when these variables appear superficially to be unrelated to residuals. In contrast, there are variables for which the response of a steel mill to large changes in those variables is so small as to be negligible. Thus, one important use of models of industrial and other activities is to determine the variables to which the activities are sensitive. However, it may be very difficult in any specific application to reflect the range of potential variation in an activity's discharges because of the difficulty in estimating future values of those sensitive variables.

5. Activities of individual economic units can be successfully modeled quantitatively. Such models can help REQM managers to make decisions on such matters as: the likely responses to charges and restrictions on the discharge of residuals to the environment; likely responses to changes in the prices, both absolute and relative, of various factor inputs; and how these responses differ between existing and new activities. For example, the steel and the pulp and paper studies demonstrated that production technology and product specifications can limit the amount of secondary materials that can be used; the petroleum refining study demonstrated the responses—and effects on residuals generation and discharge—of changes in prices of water and heat used as inputs. However, the extent to which complex, quantitative modeling of individual activities is necessary is a function of the management questions being asked, the importance of the activity relative to the residuals generators in the region, and the available analytical resources—time, personnel, computer, and data.

6. Overall REQM costs in a region are affected by REQM objectives and by physical measures and actions at subregional and regional scales, in addition to those measures adopted by individual activities. The following are examples:

A. Level and combination of ambient environmental quality (AEQ) standards. In general, REQM costs increase sharply as ambient environmental quality becomes more stringent, especially if the level and combination are (1) true for all media simultaneously; (2) to be achieved a very high percentage of the time; and (3) successfully implemented.

B. Specific sources where control is exercised to achieve given AEQ standards. For many reasons, residuals management costs for the same residual vary from source to source.
C. Spatial location of discharging activities. The same amount of discharge at different locations affects AEQ differently.
D. The extent of collective residuals modification activities. Combining residuals streams from several sources may enable achieving economies of scale in modifying and handling residuals—in materials and energy recovery, by-product production, and conventional treatment.
E. The extent of physical measures to increase directly the assimilative capacity of the environment. Physical measures to regulate the flow of rivers, introduce oxygen into water bodies, and build topography with solid residuals may result in lower total regional REQM costs than dependence solely on measures to reduce discharges from activities.

7. Quantitative models of regional REQM, useful for decision making, can be built for actual regions, as the work on the Lower Delaware Valley shows. Such models can provide estimates of the distributions of AEQ and of costs of residuals management for alternative, internally consistent sets of physical measures for REQM. The costs of building such sophisticated[1] models can be small relative to overall regional REQM costs and to the potential benefits (cost savings, effectiveness, equity considerations) of better informed decision making. This is particularly true with respect to: (a) investigating a wide range of physical measures for reducing discharges from individual activities and for modifying residuals in collective facilities; and (b) developing better models for predicting the responses of natural systems to changes in magnitudes, timing, and spatial locations of discharges. The latter is critical in evaluating the hypothesis that a particular set of physical measures will yield a desired change in AEQ. Of course, the degree of sophistication of the regional modeling effort justified in a particular case is a function of the questions to be answered, the analytical resources available, and existing data. For example, where there are only a few readily identifiable sources and only one or two AEQ indicators of primary concern, very little mod-

---

[1] Sophistication is defined in terms of the number of variables explicitly included in the model, the number of permissible options for each activity, and the accuracy with which the values of each variable and transfer coefficient are estimated.

eling—either of activities or natural systems—is likely to be necessary to develop at least a first-stage set of physical measures that will achieve the desired AEQ.

8. Although the Lower Delaware Valley case study was framed in terms of meeting AEQ standards of minimum regional costs *without* a priori specifying of discharge limitations for each major discharger, the analytical methodology used is relevant: (a) to estimate whether or not discharge limitations and best management practices if applied in a region will enable meeting AEQ standards; and (b) if not, to estimate the least-cost set of physical measures *in addition to* those limitations and practices which *will* enable meeting the AEQ standards.

9. Within a region, various subareas and groups in the population are exposed to different levels of AEQ. Similarly, REQM strategies impinge differently on different groups, in terms of the distribution of both changes in AEQ and incidence of REQM costs, as, for example, the Delaware cost study demonstrates.

10. Because of the public goods nature of AEQ, the choice of levels of different indicators of AEQ to be achieved and of the distribution of residuals management activities to achieve those levels is inherently a collective choice problem. Such choices are best made in a representative government setting. Defining the optimal jurisdiction for such choices is a function of political, technological, and economic considerations and of the characteristics of natural systems. Regional governments of general jurisdiction appear conceptually to be the most appropriate contexts for making collective choices about REQM. Because environmental quality represents only one of the relevant sectors reflecting the desires of society, the collective choices should be made in a context where tradeoffs among sectors are possible, for example, in a body with responsibility for all sectors.

11. In a realistic setting there are significant nonmarket tradeoffs among liquid, gaseous, and solid residuals and between material and energy residuals. Therefore, coherent decision making with respect to their management requires considering them simultaneously. For example, in the Lower Delaware Valley case, with high quality landfills, going from easy to tight ambient air quality standards costs about twice as much when ambient air quality standards are easy as when they are at the existing level. Interdependencies of these kinds become particularly significant when a serious effort is made to approach very low levels of residuals discharges.

12. Analysis and projections of the macroeconomic effects of the overall costs of the widespread installation and operation of physical measures to improve AEQ need not deter the implementation of such programs. They do emphasize, however, the need for both effectiveness and efficiency in the quest for improved AEQ. Both micro and regional scale analyses of REQM are essential to provide inputs for the macroanalyses and to provide a basis for interpreting the results of the macroanalyses.

Bibliography

Anderson, Frederick, Allen V. Kneese, Serge Taylor, Phillip Reed, and Russell Stevenson. *Environmental Improvement Through Economic Incentives* (Baltimore, Johns Hopkins University Press for Resources for the Future, 1978).

Ayres, Robert U. "Air Pollution in Cities," *Natural Resources Journal* vol. 9 (1969); also in Arthur F. Schreiber, Paul K. Gatons, and Richard B. Clemner, eds., *Economics of Urban Problems: Selected Readings* (Boston, Houghton Mifflin, 1971); and in Robert M. Irving and George B. Priddle, eds., *Crisis* (New York, Macmillan, 1971).

————. "Phosphorus Cycling in the Environment," *Study on Critical Environmental Problems,* M.I.T. Summer Study (July 1970).

————. "A Materials-Process-Model," in Allen V. Kneese and Blair T. Bower, eds., *Environmental Quality Analysis: Theory and Method in the Social Sciences* (Baltimore, Johns Hopkins University Press for Resources for the Future, 1972).

————. *Resource, Environment and Economics* (New York, Wiley, 1978).

————. *Futures Uncertain: Challenge to Decision-Makers* (New York, Wiley, forthcoming).

————, and James Cummings-Saxton. "The Materials Process Product Model: Theory and Applications," in William A. Vogely, ed., *Mineral Materials Modeling: A State-of-the-Art Review* (Baltimore, Johns Hopkins University Press for Resources for the Future, 1975) pp. 178–244.

————, and Ivars Gutmanis. "Technological Change, Pollution and Treatment Cost Coefficients in Input–Output Analysis," in Ronald G. Ridker, ed., *Population, Resources, and the Environment,* vol. 3 of The Commission on Population Growth and the American Future's Research Reports (Washington, D.C., GPO, 1972).

————, and Allen V. Kneese. "Environmental Pollution," *Federal Programs for the Development of Human Resources,* vol. 2 (Washington, D.C., GPO, 1968).

————, and ————. "Pollution and Environmental Quality," in Harvey S. Perloff, ed., *The Quality of the Urban Environment: Essays on "New Resources" in an Urban Age* (Baltimore, Johns Hopkins University Press for Resources for the Future, 1969).

————, and ————. "Production, Consumption, and Externalities," *American Economic Review* vol. 59, no. 3 (June 1969); also in P. Dalz and R. H.

Pentell, eds., *Bench-Mark Papers in Electrical Engineering and Computer Science* (Stroudsburg, Pa., Dowden, Hutchinson, and Ross, 1974); and in Bobbs-Merrill Reprint Series (New York, 1974) (also Resources for the Future Reprint 76).

————, and ————. "Economic and Ecological Effects of a Stationary Economy," *Annual Review of Ecology and Systematics* vol. 2 (April 1971) (Resources for the Future Reprint 99).

————, and ————. "The Sustainable Economy," in Martin Pfaff, ed., *Frontiers in Social Thought—Essays in Honor of Kenneth E. Boulding* (Amsterdam, North-Holland, distributed in the United States by American Elsevier, New York, 1976).

————, M. O. Stern, and James Cummings-Saxton. "Tax Strategies for Industrial Pollution Abatement," in Institute of Electrical and Electronic Engineers, *IEEE Transactions on Systems, Man, and Cybernetics,* December 1972.

Basta, Daniel J., and Blair T. Bower. "Point and Nonpoint Sources of Degradable and Suspended Solids: Impacts on Water Quality Management," *Journal of Soil and Water Conservation* vol. 31, no. 6 (1976) pp. 252–259.

————, James L. Lounsbury, and Blair T. Bower. *Analysis for Residuals–Environmental Quality Management,* Research Paper R-11 (Washington, D.C., Resources for the Future, 1978).

Bower, Blair T. "Residuals and Environmental Management," *Journal of the American Institute of Planners* vol. 37, no. 4 (1971).

————. "Studies of Residuals Management in Industry," in Edwin S. Mills, ed., *Economic Analysis of Environment Problems* (New York, Columbia University Press for the National Bureau of Economic Research and Resources for the Future, 1975).

————, ed. *Regional Residuals–Environmental Quality Management,* Research Paper R-7 (Washington, D.C., Resources for the Future, 1977).

————. "Water Resources Management and the Choice of Technology," *Natural Resources Forum* vol. 1, no. 2 (1977) pp. 119–129.

————. "Economic Dimensions of Waste Recycling and Re-use: Some Definitions, Facts and Issues," in David W. Pearce and Ingo Walter, eds., *Resource Conservation: Social and Economic Dimensions of Recycling* (New York, University Press, 1978).

————. "Effluent Charges as an Incentive Mechanism for Improving and Maintaining Ambient Water Quality," *Paying for Pollution, Water Quality and Effluent Charges,* Proceedings of the Conservation Foundation, Washington, D.C., 1978.

————, and Anne M. Blackburn. *Dollars and Sense, A Report on Water Quality Management in the Washington Metropolitan Area* (Bethesda, Md., Interstate Commission on the Potomac River Basin, 1975).

————, and Samuel P. Mauch. "Regional Residuals–Environmental Quality Management Modelling (RREQMM): Criteria for an Optimal Planning

Effort in the Real World," *Journal of Environmental Management* vol. 4, no. 3 (1976) pp. 275–292 (Resources for the Future Reprint 132).

————— and Walter O. Spofford, Jr. "Environmental Quality Management," *Natural Resources Journal* vol. 10, no. 4 (1970).

—————, Charles N. Ehler, and Allen V. Kneese. "Incentives for Managing the Environment," *Environmental Science and Technology* vol. 11, no. 3 (1977) pp. 250–254.

—————, George O. G. Löf, and William M. Hearon. "Residuals Generation in the Pulp and Paper Industry," *Natural Resources Journal* vol. 11, no. 4 (1971) pp. 605–623.

—————, —————, and —————. "Residuals in Manufacture of Paper," *Journal of the Environmental Engineering Division,* American Society of Civil Engineers, vol. 99, no. EE1 (February 1973) pp. 1–16.

Brown, F. Lee, and A. O. Lebeck. *Cars, Cans, and Dumps: Solutions for Rural Residuals* (Baltimore, Johns Hopkins University Press for Resources for the Future, 1976).

Burton, Ian, and Andris Auliciems. "Trends in Smoke Concentrations Before and After the Clean Air Act of 1956," *Atmospheric Environment* vol. 7 (1973) pp. 1063–1070.

—————, Robert W. Kates, and Anne Kirkby. "Public Response to a Successful Air Pollution Control Programme," in James A. Taylor, ed., *Climatic Resources and Economic Activity* (Newton Abbott (England), David and Charles, 1974) pp. 173–191.

—————, Robert W. Kates, and Gilbert F. White. Chapter 3 in Ian Burton and Robert W. Kates, *The Environment as Hazard* (New York, Oxford University Press, 1978).

—————, David E. Kromm, F. Probald, and Geoffrey Wall. "An International Comparison of Response to Air Pollution," *Journal of Environmental Management* vol. 1 (1973) pp. 363–375.

—————, Douglas Billingsley, Mark Blacksell, and Geoffrey Wall. "A Case Study of Successful Pollution Control Legislation in the United Kingdom," in W. P. Adams and F. M. Helleiner, eds. *Proceedings of the International Geographical Union Congress* (Toronto, University of Toronto Press, 1972).

Carter, Luther J. *The Florida Experience: Land and Water Policy in a Growth State* (Baltimore, Johns Hopkins University Press for Resources for the Future, 1975).

Cochran, Thomas B. *The Liquid Metal Fast Breeder Reactor: An Environmental and Economic Critique* (Baltimore, Johns Hopkins University Press for Resources for the Future, 1974).

Craine, Lyle E. *Water Management Innovations in England* (Baltimore, Johns Hopkins University Press for Resources for the Future, 1969).

—————. "Institutions for Managing Lakes and Bays," *Natural Resources Journal* (July 1971).

————. "Allocating Coastal Resources: Trade-Off and Rationing Process," in Boskwick H. Ketchum, *The Waters Edge: Critical Problems of the Coastal Zone* (Cambridge, Mass., M.I.T. Press, 1972).

d'Arge, Ralph C. "The Economics of Controlling Automotive Emissions," *California Air Environment* vol. 1, no. 7 (July–September 1970).
————. "Essay on Economic Growth and Environmental Quality," *Swedish Journal of Economics* (March 1971); also in Peter Bohm and Allen V. Kneese, eds., *Economics of Environment: Papers from Four Nations* (New York, St. Martin's Press, 1972); and in Rendigs Fels, ed., *Readings in Advanced Economics* (Homewood, Ill., Richard D. Irwin for the American Economic Association, 1974).
————. "International Trade, Environmental Quality and International Controls: Some Empirical Estimates," in Allen V. Kneese, S. E. Rolfe, and J. W. Harned, eds., *Managing the Environment* (New York, Praeger, 1971).
————. "Direct and Indirect Environmental Control Strategies," in J. Dyckman and W. Landheer, *Managing the Urban Environment,* Proceedings of the Netherlands Universities Foundation for International Cooperation, Noordwijk, Netherlands, 1972.
————. "Economic Growth and the Natural Environment," in Allen V. Kneese and Blair T. Bower, eds., *Environmental Quality Analysis: Theory and Method in the Social Sciences* (Baltimore, Johns Hopkins University Press for Resources for the Future, 1972).
————. "Economic Policies, Environmental Problems, and Land Use: A Discussion of Some Issues and Strategies in Research," in D. McAllister, ed., *Environment: A New Focus for Land Use Planning* (Washington, D.C., National Science Foundation, 1973).
————. "Observations on the Economics of Transnational Externalities," in H. Smets, *Problems of Transfrontier Pollution* (Paris, Organisation for Economic Co-operation and Development, 1973).
————. "On the Economics of Transfrontier Pollution," *International Portfolio Perspectives* (Washington, D.C., United States Information Service, 1974).
————. "Processes, Interventions, Indicators, and Models," in Forest Stearns and Tom Montag, eds., *The Urban Ecosystem: A Holistic Approach* (New York, Halsted Press, 1974).
————, ed. *Economic and Social Measures of Biologic and Climatic Change,* Climatic Impacts Assessment Project (CIAP) Monograph 6 (Washington, D.C., U.S. Department of Transportation, September 1975) p. 1200.
————. "Environmental Policy Costs: Definition, Measurement and Conjecture," in Henry M. Peskin and Eugene P. Seskin, eds., *Cost–Benefit Analysis and Water Pollution Policy* (Washington, D.C., The Urban Institute, 1975).
————. "The Role of Economic Analysis," in L. E. Coate and P. Bonner, eds., *Regional Environmental Management: Selected Proceedings of the National Conference* (New York, Wiley, 1975).

————. "Transnational Externalities and Environmental Quality," in E. Mills, ed., *Economic Analysis of Environmental Problems* (New York, Columbia University Press for the National Bureau of Economic Research and Resources for the Future, 1975).

————. "Economic Analysis of Pollution from Stratospheric Flight: A Preliminary Summing Up," in A. Broderick and T. Hard, eds., *Climatic Impact Assessment,* 4th Annual Scientific Report (Boston, U.S. Department of Transportation, 1976).

————. "Research in an Accelerating Human Environment: The Problem in Outline," in Kendall Noll, ed., *Research Needs to Anticipate Environmental Impacts of Changing Resource Usages* (Stanford, Calif., Stanford Research Institute, 1976).

————. "Transfrontier Pollution: Some Issues on Regulation," in Ingo Walter, ed., *Studies in International Environmental Economics* (New York, Wiley, 1976).

————. "Managing the Global Commons: Economic Strategies for the 1980's," in *International Perspectives for the 1980's* (New York, Council on Foreign Relations, 1977).

————, and E. K. Hunt. "Environmental Pollution, Externalities and Conventional Economic Wisdom: A Critique," *Environmental Affairs* (Spring 1971).

————, and ————. "Economic Orthodoxy and Externalities Revisited," *Environmental Affairs* (November 1971).

————, and ————. "Lemmings and Other Acquisitive Animals: Proposition on Consumption," *Journal of Economic Issues* vol. 7 (May 1973).

————, and Allen V. Kneese. "Environmental Quality and International Trade," *International Organization* vol. 26, no. 2 (Spring 1972); also in D. A. Kay, and E. B. Skolnikoff, eds., *World Eco-Crisis: International Organizations in Response* (Madison, Wisc., University of Wisconsin Press, 1972) (Resources for the Future Reprint 102).

————, and K. C. Kogiku. "Economic Growth and the Natural Environment," *Review of Economic Studies* (January 1973).

————, and William Schulze. "The Coase Proposition, Information, and Long-Run Equilibrium," *American Economic Review* vol. 64 (September 1974).

————, and James Wilen. "Governmental Control of Externalities; or the Prey Eats the Predator," *Journal of Economic Issues* (May 1974).

————, Terry N. Clark, and O. Bubik. "Controlling Automotive Emissions: Methodology and Some Preliminary Tests," *Project Clean Air Research Reports* (Berkeley, Calif., University of California, July 1970).

————, L. Eubanks, and J. Barrington. "Pollution of the Stratosphere: Toward a Dynamic Model for Regulation," in *Environmental Monitoring* (Washington, D.C., Environmental Protection Agency, 1976).

————, Ingo Walter, and Gary Hufbauer. "International Implications of Environmental Quality Control and Basic Materials Policy," in National

Academy of Sciences, *International Trade, Environmental Quality and Basic Materials Policy* (Washington, D.C., NAS, 1972).

Delson, Jerome K. "Clean Use of Coal at the Hadera Power Station," *Journal of the Association of Engineers and Architects in Israel,* (Israel) (January 1975) pp. 44–51.

Freeman, A. Myrick III. "Pollution Tax," *New Republic* vol. 162, no. 5 (June 20, 1970).

————. "The Distribution of Environmental Quality," in Allen V. Kneese and Blair T. Bower, eds., *Environmental Quality Analysis: Theory and Method in the Social Sciences* (Baltimore, Johns Hopkins University Press for Resources for the Future, 1972).

————, and Charles J. Cicchetti. "Option Demand and Consumer Surplus: Further Comment," *Quarterly Journal of Economics* vol. 85, no. 3 (August 1971).

————, and Robert H. Haveman. "Benefit–Cost Analysis and Multiple Objectives: Current Issues in Water Resource Planning," *Water Resources Research* vol. 6, no. 6 (December 1970).

————, and ————. "Water Pollution Control, River Basin Authorities and Economic Incentives: Some Current Policy Issues," *Public Policy* vol. 19, no. 1 (Winter 1971).

————, John V. Krutilla, Charles J. Cicchetti, and Clifford S. Russell. "Observations on the Economics of Irreplaceable Assets," in Allen V. Kneese and Blair T. Bower, eds., *Environmental Quality Analysis: Theory and Method in the Social Sciences* (Baltimore, Johns Hopkins University Press for Resources for the Future, 1972).

Frisken, William R. *The Atmospheric Environment of Cities,* Climatological Studies, no. 25 (Toronto, Information Canada, 1973).

————. *The Atmospheric Environment* (Baltimore, Johns Hopkins University Press for Resources for the Future, 1974).

Gianessi, Leonard, Henry M. Peskin, and Edward Wolff. "The Distributional Implications of National Air Pollution Damage Estimates," in F. Thomas Juster, *The Distribution of Economic Well-Being* (Cambridge, Mass., Ballinger for the National Bureau of Economic Research, 1977) (Resources for the Future Reprint 150).

Haefele, Edwin T. "Coalitions, Minority Representation, and Vote-Trading Probabilities," *Public Choice* (Spring 1970).

————. "A Utility Theory of Representative Government," *American Economic Review* vol. 61 (June 1971).

————. *Representative Government and Environmental Management* (Baltimore, Johns Hopkins University Press for Resources for the Future, 1973).

————. *Nuclear Merchant Ships* (Washington, D.C., Maritime Transportation Research Board, National Academy of Sciences, 1974).

————, ed. *The Governance of Common Property Resources* (Baltimore, Johns Hopkins University Press for Resources for the Future, 1975).

————. "National Economic Planning in a Whig Republic," *National Economic Planning* (Washington, D.C., Chamber of Commerce of the United States, 1976).

————. "An Algorithm for Coalition Formation," *Proceedings of the International Peace Sciences Conference 1976* Ann Arbor, Mich., 1977.

————. "Towards a New Civic Calculus," in Lowdon Wingo and Alan Evans, eds., *Public Economics and the Quality of Life* (Baltimore, Johns Hopkins University Press for Resources for the Future, 1977).

Headley, Joseph Charles. "Environmental Quality and Chemical Pesticides," *Journal of Soil and Water Conservation* vol. 21, no. 4 (July–August 1966).

————. "Estimating the Productivity of Agricultural Pesticides," *American Journal of Agricultural Economics* vol. 50, no. 1 (February 1968).

————. "The Distribution of Costs and Benefits from Environmental Protection," *Proceedings of the Western Agricultural Economics Association* (1970).

————. "The Economics of Agricultural Pest Control," in Ray F. Smith and others, eds., *Annual Review of Entomology* (1972).

————. "Economics of Environmental Quality," *Journal of Environmental Quality* vol. 1, no. 4 (October–December 1972).

————. "Environmental Quality and the Economics of Agricultural Pest Control," *European and Mediterranean Organization for the Protection of Plants Bulletin* vol. 3, no. 3 (1973).

————. "The Economics of Pest Management," in William H. Luckmann and Robert H. Metcalf, eds., *Introduction to Insect Pest Management* (New York, Wiley, 1975).

————. "Policies for Externalities Extending to Resources and Environment," in Earl O. Heady and Larry R. Whiting, eds., *Externalities in the Transformation of Agriculture* (Ames, Iowa, Iowa State University of Science and Technology Press, 1975).

————, and Allen V. Kneese. "Economic Implications of Pesticide Use," *Annals of the New York Academy of Sciences* vol. 160 (June 1969).

————, and C. Robert Taylor. "Insecticide Resistance and the Evaluation of Control Strategies for an Insect Population," *The Canadian Entomologist,* (March 1975).

————, Max Langham, and W. F. Edwards. "Pesticide Residues and Environmental Economics," *Natural Resources Journal* vol. 10, no. 4 (October 1970).

Howe, Charles W., and Blair T. Bower. "Policies for Efficient Regional Water Management," *Proceedings of the American Society of Civil Engineers* vol. 96, IR 4, (1970).

James, Ivan C. II, Blair T. Bower, and Nicholas C. Matalas. "Relative Importance of Variables in Water Resources Planning," *Water Resources Research* vol. 5, no. 6 (1969).

Kneese, Allen V. "La Calidad del Agua y el Desarrollo Economico," [The Quality of Water and Economic Development], *Boletin de la Oficina Sanitaria Panamericana* vol. 69, no. 1 (July 1970).

―――. "Economic Responsibility in the By-Products of Production," *Annals of the American Academy of Political and Social Science* (May 1970).

―――. "Protecting Our Environment and Natural Resources in the 1970's," *The Forensic Quarterly* (May 1970).

―――. "Background for the Economic Analysis of Environmental Pollution," in Peter Bohm and Allen V. Kneese, eds., *The Economics of Environment—Papers from Four Nations* (London, Macmillan, 1971) pp. 1–24; also in *The Swedish Journal of Economics* vol. 73, no. 1 (March 1971).

―――. "Environmental Pollution: Economics and Policy," *American Economic Review* vol. 61, no. 2 (May 1971).

―――. "Strategies for Environmental Management," *Public Policy* vol. 19, no. 1 (Winter 1971).

―――. "Analysis of Environmental Pollution," in Robert Dorfman and Nancy S. Dorfman, eds., *Economics of the Environment* (New York, W. W. Norton, 1972) pp. 21–44.

―――. "Discharge Capacity of Water Ways and Effluent Charges," in Selman J. Mushkin, ed., *Public Prices for Public Products* (Washington, D.C., The Urban Institute, 1972) pp. 133–151.

―――. "The Economics of Environmental Management in the United States," in Allan V. Kneese, Sidney E. Rolfe, and Joseph W. Harned, eds., *Managing the Environment—International Economic Cooperation for Pollution Control* (New York, Praeger, 1972) pp. 3–52.

―――. "Pollution and Pricing," *American Economic Review* vol. 62, no. 5 (December 1972).

―――. *The Royer Lectures* (Berkeley, Calif., University of California, 1972).

―――. "The Faustian Bargain," in Edward W. Erickson and Leonard Waverman, eds., *The Energy Question: An International Failure of Policy,* vol. 1, *The World* (Toronto, University of Toronto Press, 1974) pp. 259–266.

―――. "The Application of Economic Analysis to the Management of Water Quality," in J. Rothenberg and Ian Heggie, eds., *Management of Water Quality and Environment* (London, Macmillan, 1975).

―――. "Costs of Water Quality Improvement, Transfer Functions, and Public Policy," in Henry Peskin and Eugene Seskin, eds., *Cost–Benefit Analysis,* (Washington, D.C., The Urban Institute, 1975).

―――. *Economics and the Environment* (London, Penguin Books, 1976).

―――. "Evaluating Intangible Damages of Electric Power Development," *Proceedings of a Workshop on the Measure of Intangible Environmental Impacts,* Electric Power Research Institute, Asilomar, Pacific Grove, Calif., August 3–6, 1976.

―――. "Implementation of Alternatives," *Simposio Sobre Ambiente, Salud y Desarrollo En Las Americas* [Symposium on Clean Environment, Health and Development in the Americas] (Washington, D.C., Pan American Health Organization, 1976).

———. "Natural Resources Policy 1975–1985," *Journal of Environmental Economics and Management* vol. 3, no. 4 (December 1976) (Resources for the Future Reprint 140).

———. "Benefit–Cost Analysis and the Atom," in Rolf Steppacher, Brigitte Zogg-Walz, and Hermann Hatzfeldt, eds., *Economics in Institutional Perspective—Essays in Honor of Professor K. William Kapp* (Lexington, Mass., D.C. Heath, 1977).

———. "Quantitative Comparison of Policy Instruments for Environmental Improvement," *Decision Making in the Environmental Protection Agency,* vol. 11b (Washington, D.C., National Academy of Sciences, 1977).

———. "A Commentary on Needed Changes in the 1970 Air Quality Act Amendments," in Ann F. Friedlaender, ed., *Approaches to Controlling Air Pollution* (Cambridge, Mass., M.I.T. Press, 1978).

——— and Ralph d'Arge. "Pervasive External Costs and the Response to Society," in *The Analysis and Evaluation of Public Expenditures: The PPB System,* Joint Economic Committee, Subcommittee on Economy in Government, 91 Cong. 1 sess., 1969 (Washington, D.C., GPO) pp. 87–114. (Resources for the Future Reprint 91).

———, and Blair T. Bower. *Managing Water Quality: Economics, Technology, Institutions.* (Baltimore, Johns Hopkins University Press for Resources for the Future, 1968) (German translation, *Die Wassergütewirtschaft,* Munich, R. Oldenbourg, 1972).

———, and ———. "Causing Offsite Costs to Be Reflected in Waste Disposal Decisions," in Robert Dorfman and Nancy S. Dorfman, eds., *Economics of the Environment* (New York, W. W. Norton, 1972) pp. 135–154.

———, and ———, eds. *Environmental Quality Analysis: Theory and Method in the Social Sciences* (Baltimore, Johns Hopkins University Press for Resources for the Future, 1972).

———, and ———. "Standards, Charges, and Equity," in Robert Dorfman and Nancy S. Dorfman, eds., *Economics of the Environment* (New York, W. W. Norton, 1972) pp. 159–170.

———, and ———. "Issues Surrounding Regional Residuals–Environmental Quality Management Modeling," in Blair T. Bower, ed., *Regional Residuals–Environmental Quality Management Modeling* (Washington, D.C., Resources for the Future, 1977) (RFF Research Paper R-7).

———, and Edwin T. Haefele. "Environmental Quality and the Optimal Jurisdiction," in Alan. B. Brown, Joseph A. Licari, and Egon Neuberger, eds., *Urban and Social Economics in Market and Planned Economies—Housing, Income, and Environment,* vol. 2 (New York, Praeger, 1974).

———, and Joseph Charles Headley. "Economic Implications of Pesticide Use," *Annals of the New York Academy of Science* vol. 160 (1969).

———, and Orris C. Herfindahl. *Quality of the Environment: An Economic Approach to Some Problems in Using Land, Water, and Air* (Baltimore,

Johns Hopkins University Press for Resources for the Future, 1965). (Japanese translation published 1972.)

————, and ————. "Measuring Social and Economic Change: Benefits and Costs of Environmental Pollution," in Milton Moss, ed., *The Measurement of Economic and Social Performance* (New York, Columbia University Press for the National Bureau of Economic Research, 1973) pp. 441–503.

————, and ————. *Economic Theory of Natural Resources* (Columbus, Ohio, Charles E. Merrill, 1974).

————, and Karl-Göran Mäler. "Bribes and Charges in Pollution Control: An Aspect of the Coase Controversy," *Natural Resources Journal* vol. 13, no. 4 (October 1973).

————, and Charles L. Schultze. *Pollution, Prices, and Public Policy.* (Washington, D.C., The Brookings Institution and Resources for the Future, 1975) (Published in Spanish by Marymar Ediciones, Buenos Aires, 1976). Also in Stuart S. Nagel, ed., *Policy Studies Review Annual* (Beverly Hills, Calif., Sage Publications, 1977).

————, Robert U. Ayres, and Ralph C. d'Arge. *Economics and the Environment—A Materials Balance Approach* (Baltimore, Johns Hopkins University Press for Resources for the Future, 1970). (Italian translation published 1972; Japanese translation published by the Tokoro Shoten Publishing Company Ltd., Tokyo, Japan, 1974.)

————, Blair T. Bower, and Charles N. Ehler. "Incentives for Managing the Environment," *Environmental Science and Technology* vol. 11, no. 3 (March 1977).

————, A. Myrick Freeman III, and Robert H. Haveman. *The Economics of Environment Policy* (New York, Wiley, 1973).

————, Paul Portney, and Jon Sonstelie. "Environmental Quality, Household Migration, and Collective Choice," in Edwin T. Haefele, ed., *The Governance of Common Property Resources* (Baltimore, Johns Hopkins University Press for Resources for the Future, 1975).

Lave, Lester Bernard. "Air Pollution Damage," in Allen V. Kneese and Blair T. Bower, eds., *Environmental Quality Analysis: Theory and Method in the Social Sciences* (Baltimore, Johns Hopkins University Press for Resources for the Future, 1971) pp. 213–242; also in W. Niskanen and others, eds., *1972 Benefit–Cost and Policy Annual* (Chicago, Aldine, 1972).

————. "The Economic Costs of Air Pollution," in Frank C. Emerson, ed., *The Economics of Environmental Problems* (Ann Arbor, Mich., University of Michigan, 1973) pp. 19–39.

————. "Economic Implications of Trace Contaminants in the Air," in Andre F. LeRoy, ed., *Trace Contaminants in the Environment,* American Institute of Chemical Engineers Symposium Series no. 149, vol. 71 (1975) pp. 47–53.

————. "Coal or Nuclear: The Unintended Consequences of Electricity Generators," in *California Energy: The Economic Factors* (San Francisco, Federal Reserve Bank of San Francisco, 1976) pp. 171–182.

————. "Health Effects of Electricity Generation from Coal, Oil, and Nuclear Fuel," in H. Ashley, R. L. Rudman, and C. G. Whipple, eds., *Energy and the Environment—A Risk–Benefit Approach* (Elmsford, N.Y., Pergamon Press, 1976).

————, and L. C. Freeburg. "Health Costs to the Consumer per Megawatt-Hour of Electricity," in A. Finkel, ed., *Energy, the Environment, and Human Health,* (Chicago, American Medical Association, 1974) pp. 229–236.

————, and Eugene P. Seskin. "Air Pollution and Human Health," *Science* vol. 169 (1970) pp. 723–733; also in N. Dorfman and R. Dorfman, eds., *Economics of the Environment: Selected Readings* (New York, Norton, 1973); in L. Jaffe and L. Tribe, eds., *Casebook on Environmental Protection* (Chicago, Bracton Press, 1971); in *Biology* (New York, Druskin, 1971); in D. Thompson, *The Economics of Environmental Protection* (Cambridge, Mass., Harvard University Press, 1973); in S. Williamson, *Fundamentals of Air Pollution* (Reading, Mass., Addison Wesley, 1972); in M. Lohman, *Social Costs* (Munich, Carl Hanser Verlag, 1972); and in part A. Montagu, ed., *The Endangered Environment* (Pennsauken, N.J., Auerbach, 1972).

————, and ————. "Economics and the Environment: A Review," *Journal of the American Institute of Planners* vol. 37, no. 6 (November 1971) pp. 431–432.

————, and ————. "Health and Air Pollution: The Effect of Occupation Mix," *Swedish Journal of Economics* vol. 73, no. 1 (March 1971) pp. 76–95; also in Peter Bohm and Allen V. Kneese, eds., *The Economics of the Environment* (New York, St. Martins Press, 1972).

————, and ————. "Air Pollution, Climate and Home Heating: Their Effects on U.S. Mortality Rates," *American Journal of Public Health* vol. 62, no. 7 (July 1972) pp. 909–916.

————, and ————. "An Analysis of the Association Between U.S. Mortality and Air Pollution," *Journal of the American Statistical Association* (June 1973).

————, and ————. "Acute Relationships Among Daily Mortality, Air Pollution, and Climate," in Otto Davis, Edwin S. Mills, and G. Rothenberg, eds., *The Economics of the Environment* (New York, National Bureau of Economic Research, 1974).

————, and ————. "Does Air Pollution Shorten Lives?" in John Pratt, ed., *Statistical and Mathematical Aspects of Pollution Problems* (New York, Marcel Dekker, 1974) pp. 223–244; also in *Proceedings of the Second Annual Research Conference of the Inter-University Committee on Urban Economics* (Chicago, University of Chicago, 1970) pp. 293–328; and in W. Fairley and F. Mosteller, eds., *Statistics and Public Policy* (Reading, Mass., Addison-Wesley, 1977).

———, ———. *Air Pollution and Human Health* (Baltimore, Johns Hopkins University Press for Resources for the Future, 1977).

Löf, George O. G. "Economic Considerations in Thermal Discharge to Streams," *Proceedings of the Fifth Annual Environmental Health Research Symposium* (Albany, N.Y., State Department of Health, 1968) p. 73.

———, and Allen V. Kneese. *The Economics of Water Utilization in the Beet Sugar Industry* (Baltimore, Johns Hopkins University Press for Resources for the Future, 1968).

———, and J. C. Ward. "Economic Considerations in Thermal Discharge to Streams," in F. L. Parker and P. A. Krenkel, eds., *Engineering Aspects of Thermal Pollution*, Proceedings of the Thermal Pollution Symposium August 1968 (Nashville, Tenn., Vanderbilt University Press, 1969); also published as "Economics of Thermal Discharges," *Industrial Water Engineering* vol. 7, no. 1 (January 1970) p. 12.

———, and ———. "Economics of Thermal Pollution Control," *Journal of the Water Pollution Control Federation* vol. 42 (December 1970) p. 2102 (Resources for the Future Reprint 91).

———, Blair T. Bower, and W. M. Hearon. "Integrated Residuals Management in the Pulp and Paper Industry," in press in a publication of the American Institute of Chemical Engineers.

———, ———, and ———. "Residuals Generation in the Pulp and Paper Industry," in press in a publication of the American Society of Civil Engineers.

———, ———, and ———. "Residuals Management in Pulp and Paper Manufacture," in Kaghan, ed., *Forest Products and the Environment*, American Institute of Chemical Engineers Symposium Series, No. 133, vol. 69 (1973) p. 141.

Loucks, Daniel P. "Residuals–Environmental Quality Management: A Framework for Policy Analysis," *Journal of the Water Pollution Control Federation* vol. 42, no. 6 (June 1970).

———, and H. D. Jacoby. "The Combined Use of Optimization and Simulation Models in River Basin Planning," *Water Resources Research* vol. 8, no. 6 (December 1972); also in chapter 17, R. de Neufville and D. H. Marks, eds., *Systems Planning and Design: Case Studies in Modelling; Optimization, and Evaluation* (Engelwood Cliffs, N. J., Prentice-Hall, 1974).

———, and ———. "Flow Regulation for Water Quality Management," in R. Dorfman, H. D. Jacoby, and H. A. Thomas, Jr., eds., *Models for Managing Regional Water Quality* (Cambridge, Mass., Harvard University Press, 1972).

———, Blair T. Bower, and Walter O. Spofford, Jr. "Environmental Noise Management," *Journal of the Environmental Engineering Division, American Society of Civil Engineers* vol. 99, no. EE6 (December 1973) (Resources for the Future Reprint 113).

Luken, Ralph A. *Development Versus Preservation of Wetland Resources* (New York, Praeger, 1976).

————, and Edward Pechan. "How Much Pollution Control for What Price?" *Journal of Soil and Water Conservation* (November/December 1976).

————, and ————. *Water Pollution Control: Assessing the Impacts and Costs of Environmental Standards* (New York, Praeger, 1977).

Mäler, Karl-Göran. "Environmental Control and Economy," *Skandinaviska Banken Quarterly Review* vol. 3 (1971) pp. 63–68.

————. *Environmental Economics: A Theoretical Inquiry* (Baltimore, Johns Hopkins University Press for Resources for the Future, 1974).

Miller, Delbert C. "The Allocation of Priorities to Urban and Environmental Problems by Powerful Leaders and Organizations," in W. Burch, N. Cheek, and L. Taylor, eds., *Social Behavior, Natural Resources, and the Environment* (New York, Harper & Row, 1972).

————. "Power Structure Studies and Environmental Management," in Allen V. Kneese and Blair T. Bower, eds., *Environmental Quality Analysis: Theory and Method in the Social Sciences* (Baltimore, Johns Hopkins University Press for Resources for the Future, 1972).

————. *Top Leadership and Organizational Power in Megalopolis, Environmental, Ecology and Urban Organization* (New York, Wiley, Interscience, 1975).

————. "Systems Characteristics of Megalopolis," in Roland L. Warren, *New Perspectives on the American Community*, 3rd ed. (Skokie, Ill., Rand McNally, 1977) pp. 135–144.

Olson, Mancur. "The Economics of Integrative Systems," in Bernhard Kulp and Wolfgang Stutzel, eds., *Beiträge Zu Einer Theorie der Sozialpolitik, Festschrift für Elisabeth Leifmann-Keil* (Berlin, Dunken and Jumbolt, 1973) pp. 31–42.

————. "Evaluating Performance in the Public Sector," in Milton Moss, ed., *The Measurement of Economic and Social Performance*, Studies in Income and Wealth, vol. 38 (New York, Columbia University Press for the National Bureau of Economic Research, 1973) pp. 355–384.

————. "The Priority of Public Problems," in Robin Harris, ed., *The Corporate Society* (London, Macmillan, 1974) pp. 294–336.

————. "Assessing the 'Quality of Life'," *Contemporary Review*, vol. 227 (December 1975) pp. 321–324.

————. "Ignorance and Uncertainty," *Proceedings of the Society for General Systems Theory* (1975).

————. "Preliminary Thoughts about the Causes of Harmony and Conflict," in Robert D. Leiter and Gerald Sirkin, eds., *Economics of Public Choice* (New York, Cyrco Press for City College of the City University of New York, 1975) pp. 160–167.

————. "Cost–Benefit Analysis, Statistical Decision Theory, and Environmental Policy," *Proceedings of the Philosophy of Science Association*, vol. 2 (1976) pp. 372–394.

————. "The Political Economy of Comparative Growth Rates," *U.S. Economic Growth from 1976 to 1986.* Hearings before the Joint Economic Committee, U.S. Cong., November 10, 1976 (Washington, D.C., GPO).

————. "The Treatment of Externalities in National Income Statistics," in Lowdon Wingo and Alan Evans, eds., *Public Economics and the Quality of Life* (Baltimore, Johns Hopkins University Press for Resources for the Future and the Centre for Environmental Studies, 1977) pp. 219–249.

————, and Hans H. Landsberg, eds. *The No Growth Society* (New York, W. W. Norton, 1974).

————, and Richard Zeckhauser. "The Efficient Production of External Economies," *American Economic Review* vol. 60 (June 1970) pp. 512–517.

Page, Talbot R. "Economics of Recycling," *Resource Conservation, Resource Recovery, and Solid Waste Disposal.* Studies prepared by the Committee on Public Works, U.S. Senate (Washington, D.C., Library of Congress, 1973); reprinted in *Scrap Age* (April 1974) pp. 183–222.

————. "Failure of Bribes and Standards for Pollution Abatement," *Natural Resource Journal,* Symposium on the Coase Theorem (October 1973).

————. "How Much Recycling is Enough?" *Cycling and Control of Metals,* proceedings of an Environmental Protection Agency conference, October 31–November 2, 1972 (Washington, D.C., EPA, February 1973).

————. "Pollution Affecting Producers in an Input–Output Context," *Institute of Electrical and Electronics Engineers Transactions on Systems, Man and Cybernetics* (November 1973).

————. "Recycling, Taxes, and Conservation," *National Parks and Conservation Magazine* (January 1973).

————. "Materials Policy," *The Need for a National Materials Policy.* Hearings before the Panel on Materials Policy of the Subcommittee on Environmental Pollution of the Committee on Public Works, U.S. Senate, 93 Cong., 2 sess., June 11, 1974, Serial No. 93-H47 (Washington, D.C., GPO).

————. *Conservation and Economic Efficiency: An Approach to Materials Policy* (Baltimore, Johns Hopkins University Press for Resources for the Future, 1977).

————. "Discounting and Intergenerational Equity," *Futures* (October 1977).

————. "Equitable Use of the Resource Base," *Environment and Planning* vol. 9 (January 1977) (Resources for the Future Reprint 144).

————. "Intertemporal and International Aspects of Virgin Materials Taxes," in David Pearce and Ingo Walter, eds., *Resource Conservation: Social and Economic Dimensions of Recycling* (New York, New York University Press, 1977).

————, and John Ferejohn. "Externalities as Commodities: Comment," *American Economic Review* vol. 64 (June 1974).

————, and ————. "On the Foundations of Intertemporal Choice," *American Journal of Agricultural Economics* vol. 60, no. 2 (May 1978) (Resources for the Future Reprint 152).

————, and John V. Krutilla. "Towards a Responsible Energy Policy," *Policy Analysis* (February 1975); published as "Energy Policy from an Environmental Perspective," in Robert J. Kalter and William A. Vogely, eds., *Energy Supply and Government Policy* (Ithaca, N.Y., Cornell University Press, 1976).

————, and Mark F. Sharefkin. "Industry Influence on Environmental Decision Making," in Edwin T. Haefele, ed., *The Governance of Common Property Resources* (Baltimore, Johns Hopkins University Press for Resources for the Future, 1975).

————, William Thomas, Nicholas Ashford, and Edward Burger. "Working Paper on Equity," *Decision Making for Regulating Chemicals in the Environment* (Washington, D.C., National Academy of Sciences, 1975).

Peskin, Henry M. *National Accounting and the Environment,* Artikler 50 (Oslo, Statistisk Sentralbyrå, 1972).

————. "National Accounting and the Environment: A Progress Report," *Social Indicators Research* (Fall 1975); also in abridged form in George Rohrlich, ed., *Environmental Management* (Cambridge, Mass., Ballinger, 1976).

————. "A National Accounting Framework for Environmental Assets," *Journal of Environmental Economics and Management,* vol. 2 (1976) pp. 255–262.

————. "Environmental Policy and the Distribution of Benefits and Costs," in Paul R. Portney, ed., A. Myrick Freeman III, Robert H. Haveman, Henry M. Peskin, Eugene P. Seskin, and V. Kerry Smith, *Current Issues in U.S. Environmental Policy* (Baltimore, Johns Hopkins University Press for Resources for the Future, 1978).

————, and Janice Peskin. "The Valuation of Nonmarket Activities in Income Accounting," *Review of Income and Wealth* (March 1978).

————, Leonard P. Gianessi, and Edward Wolff. "The Distributional Effects of the Uniform Air Pollution Policy in the United States," Discussion Paper D-5 (Washington, D.C., Resources for the Future, March 1977); *Quarterly Journal of Economics* (in press).

Ridker, Ronald G. "Population and Pollution in the United States," *Science* vol. 176, no. 4039 (July 9, 1972) pp. 1085–1090.

————. Population, Resources, and the Environment, vol. 3 of *Research Reports of the Commission on Population Growth and the American Future* (Washington, D.C., GPO, 1972).

————. "Impact of Population Growth on Resources and the Environment," in Charles F. Westoff and coauthors, *Toward the End of Growth,* (Englewood Cliffs, N.J., Prentice-Hall, 1973).

————. "To Grow or Not to Grow: That is not the Relevant Question," *Science* vol. 182 (December 28, 1973) pp. 1315–1318.

————. "Resources and Amenity Implications of Changes in Population Growth Rates," *Papers and Proceedings of the American Economic Review* vol. 64, no. 2 (May 1974).

————. "Population Growth, Economic Growth, and the Environment in the United States," in Ansley J. Coale, ed., *Economic Factors in Population Growth* (London, Macmillan, 1976).

————. "Materials Supply Problems from the Viewpoint of Resource-Poor, Less Developed Nations," *Materials and Society* vol. 1 (Elmsford, New York, Pergamon Press, 1977) pp. 23–26.

————. "The Effects of Slowing Population Growth on Long-Run Economic Growth in the United States During the Next Half Century," in Thomas J. Espenshade and William J. Serow, eds., *The Economic Consequences of Slowing Population Growth* (New York, Academic Press, 1978) (Resources for the Future Reprint 160).

————, and Pierre R. Crosson. "Resource, Environment and Population," in Warren Robinson, ed., *Population and Development Planning* (New York, Population Council, 1975).

————, and Joseph L. Fisher. "Effects of Population Growth on Resource Availability and Environmental Quality (and Vice Versa)," *The American Economic Review* vol. 63, no. 2 (May 1973) pp. 79–87.

————; and Allen V. Kneese. Review of *The Limits to Growth*, by Donella H. Meadows, Dennis L. Meadows, Jorgen Randers, and William W. Behrens, *The Washington Post*, March 2, 1972.

————, William D. Watson, Jr., and Adele Shapanka. "Economic, Energy, and Environmental Consequences of Alternative Energy Regimes: An Application of the RFF/SEAS Modeling System," in Charles J. Hitch, ed., *Modeling Energy-Economic Interactions: Five Approaches,* Research Paper R-5 (Washington, D.C., Resources for the Future, 1977).

Russell, Clifford S. "Models for the Investigation of Industrial Response to Residuals Management Actions," *Swedish Journal of Economics* vol. 73, no. 1 (March 1971) pp. 134–156.

————. "Application of Micro Economic Models to Regional Environmental Quality Management," *American Economic Review Papers and Proceedings* vol. 63, no. 2 (May 1973) pp. 236–243.

————. "The Municipal Evaluation of Regional Water Quality Management Proposals," in Robert Dorfman, H. D. Jacoby, and H. A. Thomas, eds., *Models for Managing Regional Water Quality* (Cambridge, Mass., Harvard University Press, 1973).

————. *Residuals Management in Industry: A Case Study of Petroleum Refining* (Baltimore, Johns Hopkins University Press for Resources for the Future, 1973).

————. "Restraining Demand by Pricing Water Withdrawals and Waste Disposal," in B. M. Fennell and R. D. Hey, eds., *The Management of Water Resources in England and Wales* (London, Saxon House, 1974).

————, ed. *Ecological Modeling in a Resource Management Framework,* Working Paper QE-1 (Washington, D.C., Resources for the Future, 1975).

————, ed. *Safe Drinking Water: Current and Future Problems,* Research Paper R-12 (Washington, D.C., Resources for the Future, 1978).

————. "Vote Trading: An Attempt at Clarification," *Political Studies* (in press).

————. "Characteristics of Effluent Charge Systems," *Policy Analysis* (in press).

————, and Allen V. Kneese. "Establishing the Scientific, Technical, and Economic Basis for Coastal Zone Management," *Coastal Zone Management Journal* vol. 1, no. 1 (1973) pp. 47–63.

————, and Hans Landsberg. "International Environmental Problems, A Taxonomy," *Science* (June 25, 1971) pp. 1307–1314.

————, and V. Kerry Smith. "The Selection of Macro-Policy Instruments Over Time," *The Australian Economic Papers* (June 1975).

————, and Walter O. Spofford, Jr., "A Quantitative Framework for Residuals Management Decisions," in Allen V. Kneese and Blair T. Bower, eds., *Environmental Quality Analysis: Theory and Method in the Social Sciences* (Baltimore: Johns Hopkins University Press for Resources for the Future, 1972).

————, and ————. "A Regional Environmental Quality Management Model: An Assessment," *Journal of Environmental Economics and Management* vol. 4, no. 2 (1977) (Resources for the Future Reprint 147).

————, and William J. Vaughan. "A Linear Programing Model of Residuals Management for Integrated Iron and Steel Production," *Journal of Environmental Economics and Management* vol. 1, no. 1 (1974) pp. 17–42.

————, and ————. *Steel Production: Processes, Products, and Residuals* (Baltimore, Johns Hopkins University Press for Resources for the Future, 1976).

————, and ————. "An Analysis of the Historical Choice Among Technologies in the U.S. Steel Industry: Contributions from a Linear Programming Model," *The Engineering Economist* vol. 22, no. 1 (1977).

————, Walter O. Spofford, Jr., and Edwin T. Haefele. "The Management of the Quality of the Environment," in J. Rothenberg and Ian G. Heggie, eds., *The Management of Water Quality and the Environment* (London, Macmillan, 1974).

————, Walter O. Spofford, Jr., and Robert A. Kelly. "Interdependencies Among Gaseous, Liquid and Solid Residuals: The Case of the Lower Delaware Valley," in Gerald J. Karaska and Craig L. Moore, eds., *Northeast Regional Science Review* vol. 5 (1975) pp. 23–35.

————, John V. Krutilla, Charles J. Cicchetti, and A. Myrick Freeman III. "Observations on the Economics of Irreplaceable Assets," in Allen V. Kneese and Blair T. Bower, eds., *Environmental Quality Analysis: Theory and Method in the Social Sciences* (Baltimore, Johns Hopkins University Press for Resources for the Future, 1972).

Sawyer, James W., Jr. *Automotive Scrap Recycling: Processes, Prices, and Prospects* (Baltimore, Johns Hopkins University Press for Resources for the Future, 1974).

————, Blair T. Bower, and George O. G. Löf. "Modeling Process Substitutions by LP and MIP," in Robert M. Thrall and others, eds., *Economic Modeling for Water Policy Evaluation*, TIMS Studies in the Management Sciences, vol. 3 (New York, North-Holland, 1976) pp. 157–178.

Sax, Joseph L. "The Public Trust Doctrine in Natural Resource Law: Effective Judicial Intervention," *Michigan Law Review* vol. 68, no. 3 (January 1970) pp. 471–566.

————. *Defending the Environment* (New York, Alfred A. Knopf, 1971).

Sewell, W. R. Derrick. "Environmental Perceptions and Attitudes of Engineers and Public Health Officials," *Environment and Behavior* (March 1971) pp. 23–59; also in J. F. Wohlwill and D. H. Carson, eds., *Environment and the Social Sciences: Perspectives and Applications* (Washington, D.C., American Psychological Association, 1972) pp. 249–264; in John H. Sims and Duane Baumann, eds., *Human Behavior and the Environment: Interactions Between Man and His Physical World.* (Chicago, Maaroufa Press, 1974) pp. 179–214; and in Andrew Porteous, Keith Attenborough, and Christopher Pollitt, eds., *Pollution: The Professionals and the Public* (Milton Keynes, The Open University, 1977) pp. 139–166.

————. "The Role of Professionals in Environmental Decision-Making," in J. T. Coppock and C. B. Wilson, eds., *Environmental Quality* (Edinburgh, Scottish Academic Press, 1974) pp. 109–131.

————, and L. R. Barr. "Evolution in the British Institutional Framework for Water Management," *Natural Resources Journal* (July 1977) pp. 395–413.

————, and ————. "Water Administration in England and Wales: Impacts of Re-organization," *Water Resources Bulletin* vol. 14, no. 2 (April 1978) pp. 337–348.

————, and Blair T. Bower, eds., *Forecasting the Demands for Water* (Ottawa, Queen's Printer, January 1969).

————, and ————. *Selecting Strategies for Air Quality Management* (Ottawa, Queen's Printer, 1970).

————, and Ian Burton, eds. *Perceptions and Attitudes in Resources Management* (Ottawa, Information Canada, 1971).

————, and Peter Harrison. "Water Reorganization in France: A Decade of Readjustment," *Water* (January 1975) pp. 44–48.

Spofford, Walter O., Jr. "Closing the Gap in Waste Management," *Environmental Science and Technology* vol. 4, no. 12 (December 1970) pp. 1108–1114.

————. "Decision Making Under Uncertainty: The Case of Carbon Dioxide Build-up in the Atmosphere," in W. H. Matthews, W. H. Kellogg, and G. D. Robinson, eds., *Man's Impact on the Climate* (Cambridge, Mass., M.I.T. Press, 1971).

————. "Residuals Management: An Overview of the Global Problems," in W. H. Matthews, F. E. Smith, and E. D. Goldberg, eds., *Man's Impact on Terrestrial and Oceanic Ecosystems* (Cambridge, Mass., M.I.T. Press, 1971).

————. "Solid Residuals Management: Some Economic Considerations," *Natural Resources Journal* vol. 11, no. 3 (July 1971) pp. 561–589.

————. "Solid Waste—The Junk We Discard," prepared for 11th Carnegie Conference on Pollution: Planning Our Environment, Pittsburgh, Pa., Carnegie-Mellon University, May 1, 1971.

————. "A Mathematical Model for Selecting Optimal Abatement Strategies in Water Quality Management," *First Symposium on Problems of Water Resources Systems* vol. 2, Karlovy Vary, Czechoslovakia, May 23–27, 1972 (Praha, 1973).

————. "Total Environmental Quality Management Models," in Rolf A. Deininger, ed., *Models for Environmental Pollution Control* (Ann Arbor, Mich., Ann Arbor Science Publishers, 1973) (Resources for the Future Reprint 130).

————. "Ecological Modeling in a Resource Management Framework: An Introduction," in Clifford S. Russell, ed., *Ecological Modeling in a Resource Management Framework,* Working Paper QE-1 (Washington, D.C., Resources for the Future, 1975).

————, and Blair T. Bower. "Environmental Quality Management," *Natural Resources Journal* vol. 10, no. 4 (October 1970) pp. 655–667.

————, and Charles N. Ehler. "Regional Residuals–Environmental Quality Management Models: Applications to EPA's Regional Management Programs," *Proceedings of the Conference on Environmental Modeling and Simulation,* Cincinnati, Ohio, April 19–22, 1976 (Washington, D.C., U.S. Environmental Protection Agency, July 1976) (EPA 600/9-76-016).

————, and Robert A. Kelly. "Application of an Ecosystem Model to Water Quality Management: The Delaware Estuary," in C. A. S. Hall and J. W. Day, Jr., eds., *Ecosystem Modeling in Theory and Practice: An Introduction with Case Histories* (New York, Wiley, 1977).

————, and Clifford S. Russell. "A Quantitative Framework for Residuals–Environmental Quality Management," in G. H. Toebes, ed., *Natural Resource Systems Models in Decision Making,* proceedings of a Water Resources Seminar, Water Resources Research Center, Purdue University, 1969.

————, and ————. "A Regional Environmental Quality Management Model: An Assessment," *Journal of Environmental Economics and Management* vol. 4, no. 2 (June 1977) pp. 89–110. (Resources for the Future Reprint 147).

————, ————, and R. A. Kelly. "Interdependencies Among Gaseous, Liquid, and Solid Residuals: The Case of the Lower Delaware Valley," *Northeast Regional Science Review* vol. 5 (November 1975).

————, ————, and ————. "Operational Problems in Large Scale Residuals Management Models." Edwin S. Mills, ed., *Economic Analysis of Environ-*

mental Problems (New York, Columbia University Press for the National Bureau of Economic Research and Resources for the Future, 1975).

————, ————, and ————. *Environmental Quality Management: An Application to the Lower Delaware Valley,* Research Paper R-1 (Washington, D.C., Resources for the Future, 1976).

————, ————, and ————. "The Lower Delaware Valley Integrated Residuals Management Model: A Summary," in Blair T. Bower, ed., *Regional Residuals–Environmental Quality Management Modeling,* Research Paper R-7 (Washington, D.C., Resources for the Future, 1977).

Index

Acid mine drainage, 44, 100

Activity models, 114, 116, 117–127, 139, 145, 146; *see also* Linear programming models

Aggregation of information: data, 201, 238, 242; external costs, 10; gaseous residuals, 219; individual preferences, 169, 170, 192–194; industrial water needs, 53; labor productivity, 280, 289–290; national accounting, 255; national cost–benefit, 268; national final demand, 35; national production, *see* Gross national product; Net national product; residuals–environmental quality management costs, 113, 220

Agricultural residuals, 268

Air quality, 207–208, 213, 219; costs, 223–225, 228, 230–231, 240, 269; Implementation Planning Program (IPP), 160, 219, 239; modeling, 209, 242, *see also* Atmospheric dispersion models; standards, 220, 221, 230, 234

Airborne residuals, 12; *see also* Gaseous residuals; Particulates; Sulfur dioxide

Algal density, 163, 209, 213, 219

Almon, Clopper, 252, 271

Ambient environmental quality (AEQ), 14, 16, 29, 32, 51, 107, 137, 142, 194, 261; constraints, 139, 153, 154, 222; costs, 140, 242, 266, 268, 286–288, 291fn; distribution, 204, 209, 212–213, 230–237,

244–245; expenditures, national accounting, 288–290; indicators, 184, 213, 303; standards, 228, 286, 288, 289, 302–303, 305, specification, 221–222; vector analysis, 211; *see also* Air quality; Water quality

Ammonia (NH_3), 92; effluent charges, 76, 78, 83; recovery, 85

Aquatic ecosystem model, 139, 143, 146, 153, 159, 162–167, 219, 242, 243–244; components, 163; data collection, 241; material transfers, 163–165; residuals discharge vector, 211; *see also* Delaware Estuary model

Arrow, Kenneth, 168–169; Arrow Paradox, 195–196; Impossibility Theorem, 168, 196, 197; Possibility Theorem, 169, 197

Assimilative capacity, 300; increasing, 7, 12, 18, 51–52, 145, 303

Atmospheric dispersion, 207

Atmospheric dispersion model, 153, 159, 160–162, 219–220, 239; residuals discharge vectors, 211

Automobile discharge, 269

Automotive scrap: battery lead, 47; steel, 34, 48, 94–100, dismantling operations, 94, 96, policies affecting use of, 99–100

Ayres, Robert U., 24

Bacteria mass, 163, 219

Basic oxygen furnace (BOF), 61, 83, 84, 85, 92

Library of Congress Cataloging in Publication Data

Kneese, Allen V.
    Environmental quality and residuals management.

    1. Environmental policy.   2. Pollution—
Economic aspects.   3. Waste products.   I. Bower,
Blair T., joint author.   II. Resources for the
Future.   III. Title.
HC79.E5K576   301.31   79-2181
ISBN 0-8018-2245-9
ISBN 0-8018-2286-6 pbk.